Political Economy and Information Capitalism in India

Technology, Globalisation and Development Series

Series Editor: **Anthony P. D'Costa,** Professor in Comparative International Development, University of Washington, USA.

The series examines technological change in the larger global context by identifying where the markets are, who is specializing in what areas of technology and why, what is the nature of global trade in technologies, and what are some of the social and political challenges posed by such change. Relationships between the OECD and rapidly growing Asian and Latin American economies are brought out by this series. At the same time, the series will also address the issue of why some regions such as Africa and the Middle East lag behind. The question of whether technology is important for development, and how various governments have targeted technology development, whether through education, skill development, or R&D also adds to the diversity of the series. The series includes comparative and regional studies, including single-country case studies of large economies such as India, China, Japan, South Korea, and Brazil. While no single approach or methodology is used exclusively, scholars within the series rely on political economy approaches that are sensitive to institutional and historical realities when analyzing specific countries/regions or technologies.

Titles include:

Govindan Parayil *(editor)*
POLITICAL ECONOMY AND INFORMATION CAPITALISM IN INDIA
Digital Divide, Development and Equity

Technology, Globalisation and Development Series.
Series Standing Order ISBN 1–4039–3591–2

You can receive future titles in this series as they are published by placing a standing order. Please contact your bookseller or, in case of difficulty, write to us at the address below with your name and address, the title of the series and one of the ISBNs quoted above.

Customer Services Department, Macmillan Distribution Ltd, Houndmills, Basingstoke, Hampshire RG21 6XS, England

Political Economy and Information Capitalism in India

Digital Divide, Development and Equity

Edited by

Govindan Parayil

palgrave
macmillan

First published 2006 by
PALGRAVE MACMILLAN
Houndmills, Basingstoke, Hampshire RG21 6XS and
175 Fifth Avenue, New York, N.Y. 10010
Companies and representatives throughout the world

PALGRAVE MACMILLAN is the global academic imprint of the Palgrave Macmillan division of St. Martin's Press, LLC and of Palgrave Macmillan Ltd. Macmillan® is a registered trademark in the United States, United Kingdom and other countries. Palgrave is a registered trademark in the European Union and other countries.

ISBN-13: 978–1–403–9–9244–4
ISBN-10: 1–403–9–9244–4

This book is printed on paper suitable for recycling and made from fully managed and sustained forest sources.

A catalogue record for this book is available from the British Library.

Library of Congress Cataloging-in-Publication Data
Political economy and information capitalism in India: digital divide, development and equity / Govindan Parayil, editor.
p. cm. – (Technology, globalization and development)
Includes bibliographical references and index.
ISBN 1–4039–9244–4 (cloth)
1. Information technology – Economic aspects – India.
2. Telecommunication – India. 3. India – Economic conditions.
4. India – Economic policy – 1991– I. Parayil, Govindan, 1955–
II. Series.
HC440.I55P65 2005
303.48′33′0954 – dc22

2005047533

10 9 8 7 6 5 4 3 2 1
15 14 13 12 11 10 09 08 07 06

Printed and bound in Great Britain by
Antony Rowe Ltd, Chippenham and Eastbourne

Contents

List of Tables

List of Figures

Acknowledgements

Anthony P. D'Costa: I would like to thank Tojo Thatchenkerry for inviting me to present an earlier version of this chapter at the National Science Foundation, Washington, DC-sponsored 'Workshop on Learning from the Indian Development Experience', held at the Indian Institute of Science, Bangalore in March 2003. Both Govindan Parayil and Janette Rawlings helped in vastly revising the paper. The usual disclaimers apply.

T. T. Sreekumar: Thanks are due to Erik Baark and Greg Felker for critical comments and suggestions that helped shape the arguments in this chapter. I wish to thank Agnus Ku and Govindan Parayil for useful discussions and comments on earlier versions of the chapter. I have benefited from comments offered by Ashwini Saith, Steve Fuller, Anthony D'Costa, Tojo Thatchenkerry, Sundar Sarukkai, Alakh N. Sharma, Subhash Bhatnagar and Narelle George during seminars and workshops where earlier versions or parts of this chapter have been presented. Thanks are also due to Satish Jha, Subbaiah Arunachalam and Ranjith Khosla for perceptive discussions, although our views on CSO initiatives might differ drastically. I thank them all. I would also like to thank K. Ramachandran for discussions and support during fieldwork. The usual disclaimers apply.

Jayan Jose Thomas: I thank the National University of Singapore for providing a research grant to carry out this study. I received excellent research assistance from Askarmon, George V. Thomas, M. Hemasekhar, J. K. Manjunath, K. Purushotham and Sebastian K. Jose while carrying out the survey. Govindan Parayil, V. K. Ramachandran, Madhura Swaminathan and Milagros Rivera provided useful suggestions and criticism on this study. I am very grateful to all of them.

Preface

The person who probably deserves the most credit for selling India's emerging capabilities in software and information technology-enabled services (ITES) to the world was the late Dewang Mehta. As President of the National Association of Software and Service Companies (NASSCOM), Mehta travelled around the globe to boost the fortunes of Indian software and ITES industry. I met Mehta in Salzburg, Austria in July 1998 at the Salzburg Seminar on 'Social and Political Implications of the Internet'. As one of the scheduled speakers at the Seminar, Mehta gave a memorable talk about the emergence of India as an IT powerhouse, peppered with wry accountants' jokes, to a largely academic audience. One of his jokes that still lingers in my mind is his quip that India already owned half of the Internet because the zero was an ancient Indian invention. Besides Dewang Mehta, other visionaries of India's software industry such as Narayana Murthy averred that India may have missed the Industrial Revolution but it cannot afford to miss the Information Revolution, and that India must leverage its emerging prowess in software to become a developed nation by 2020.

Taking stock of the impact of information and communications technologies (ICTs) on the Indian economy in 2005, one feels that, despite the dynamism of India's thriving software and ITES industry and the rhetorical flourishes of industry leaders and IT bureaucrats, India needs to do a lot more to ride the waves of ICT innovation in order to benefit the hundreds of millions of Indians eking out a living in India's vast rural hinterlands and sprawling urban slums on less than a dollar a day, rather than let the ICT Revolution benefit just the hundreds of thousands with the right cognitive skills who seem to have succeeded in the so-called knowledge economy. The chapters in this volume are intended to assess the equity implications of the ICT Revolution in India, to reflect on and to analyse the political economy of information technology and development.

Ideas and thoughts that form the core of several chapters in this volume were presented and discussed by the editor and many of the contributors at the Third International Conference of Asia Scholars (ICAS3), Singapore, August 2003; at the International Workshop on 'ICTs and Indian Development' organised by the School of Public Policy, George Mason University and the Department of Management Studies,

Indian Institute of Science, Bangalore, March 2003; at the Plenary Session on 'Innovation, Learning and Developing Countries' at the Second Globelics Conference, Beijing, October 2004; and at the International Conference on the Digital Divide: Technology and Politics in the Information Age, Hong Kong Baptist University, August 2002.

I am grateful to Jennifer Nelson, Social Sciences Editor at Palgrave Macmillan for commissioning this book project and to Anthony D'Costa, Technology, Globalisation and Development Series Editor for Palgrave Macmillan, for his encouragement and fruitful suggestions. I am also grateful to Helge Hveem, Director, Centre for Technology, Innovation and Culture, University of Oslo for his support. Without the love and support from my wife Olive, I could not have found the time and space to complete this project. Finally, I dedicate this book to my children, Sushmita and Anik. Savvy users of computers and the Internet, they are progenies of the Information Age. Perhaps the full impact of the Information Revolution will be felt in their lifetime.

G. P.
15 March 2005, Oslo

List of Contributors

C. P. Chandrasekhar is Professor at the Centre for Economic Studies and Planning, Jawaharlal Nehru University, New Delhi, India (cpc@vsnl.com).

Anthony P. D'Costa is Professor of Comparative International Development, Interdisciplinary Arts and Sciences Program, University of Washington, Tacoma, Washington, USA (dcosta@u.washington.edu).

Kyle Eischen is Associate Director of Regional and Informational Research at the Center for Global, International and Regional Studies, University of California, Santa Cruz, USA (keischen@ucsc.edu).

K. J. Joseph is a Fellow at the Centre for Development Studies, Thiruvananthapuram, Kerala, India (kjjoseph@cds.ac.in).

Govindan Parayil is Professor at the Centre for Technology, Innovation and Culture, University of Oslo, Norway (govindan.parayil@tik.uio.no).

Balaji Parthasarathy is ICICI Assistant Professor at the International Institute of Information Technology, Bangalore, India (pbalaji@iiitb.ac.in).

T. T. Sreekumar is Assistant Professor at the Communications and New Media Programme, National University of Singapore (sreekumar@nus.edu.sg).

Jayan Jose Thomas is Visiting Research Fellow at the Institute of South Asian Studies, National University of Singapore (isastjj@nus.edu.sg).

Biao Xiang is a research officer at the Centre on Migration, Policy and Society (COMPAS), University of Oxford, UK (biao.xiang@compas.ox.ac.uk).

1
Introduction: Information Capitalism

Govindan Parayil

The far-reaching social, economic and political changes due to recent advances in information and communications technologies (ICTs) in tandem with the globalisation of trade, investment, production and consumption have heralded the rise of 'information capitalism', or 'knowledge capitalism'. Information capitalism is referred to as the economic logic of post-industrialism that posits a rupture in the articulation of industrial capitalism, which was based on social labour, industrial capital, manufacturing and bulk production. Information or knowledge capitalism is not a continuation of industrial capitalism by other means. Also, it is not merely a form of hyper-industrialism or post-Fordist flexible production arrangement. It indicates a rupture of industrial capitalism and the evolution of an entirely new economic dynamics. The morphing of industrialism into informational capitalism occurred with a fundamental change in the meaning and significance attributed to knowledge as well as the changing relationship between capital, labour and knowledge. Economic changes do not take place in a political–economic vacuum. The newly evolving informational economy is a reflection of the social and political changes taking place in the world at large due to various factors, most notably the globalisation of businesses and the increasing global integration of local and national economies and markets.

Until the rise of the so-called knowledge capitalism, the industrialised world was made up of Western Europe, North America and Japan; only a handful of developing countries, especially the four Asian 'tigers', seemed to have made it to the industrial capitalist camp as fully fledged members. Of late, China has made tremendous progress as an industrial manufacturing giant whose export-led economy has been growing at breakneck speed, while neighbouring India has made only marginal strides in the area of manufacturing and export-led industrialisation to eradicate rampant poverty and deprivation. China has also performed better than India on every economic indicator.

In a totally unexpected way, recent advances in ICTs appeared to have opened up new opportunities for India as a major exporter of information technology-enabled services (ITES) and software services, thanks to economic globalisation and liberalisation of the domestic economy. Information technology-led development has spurred renewed hopes and expectations of India's becoming a major economic power. For example, India's emergence as an outsourcing 'super power', made possible by its capabilities in the software and ITES sectors, has rekindled the hope that large-scale employment possibilities would be opened up for millions of college-educated Indians. The foreign exchange windfall from the export of software and ITES is touted to power the engine of India's economy in the coming decades.

Concurrently, a resurgence of ICT fetishism and cyber-libertarian development thinking has enveloped the discourse on information revolution and socio-economic development in the world, which has been spearheaded by prominent international development institutions since the 1990s. This euphoria on the prospects of IT-led development is more rampant in India than anywhere else in the developing world. This euphoria also occurred with the gradual disappearance of various modernisation theories of development and their displacement with neo-liberal development discourse towards the end of the twentieth century. What globalisation and the emergence of the so-called 'knowledge economy' has brought about is a total erasure of such hitherto notions as agency, causality, historical memory, welfare state, social democracy and, ultimately, the very idea of progress through large-scale social transformation. In developing countries, debates on such historical legacies as colonialism, imperialism and exploitation of their resources have been turned upside down. In the evolving neo-liberal political discourse on change and development, victims have become the new aggressors. Their poverty and deprivation is blamed on their own intransigence and failure to help themselves. Although most developing countries' abject failure to eradicate poverty and deprivation cannot be blamed exclusively on geo-political problems they had inherited post-independence, and, of late, vitiated by the dizzying frenzy of the forces of globalisation, the larger political economy question of external factors that also compounded their impoverishment and economic dependence became a non-negotiable or even non-debatable topic. The radical turnaround in the terms of development debate from its geo-political dimensions of exploitation and unequal exchange practices saw the rise of a new development paradigm centred on the premise of community-based self-help enterprises that sought the intervention of markets and financial capital. The failure of states to deliver on their development promises became one of the central premises of the cyber-libertarian development model of the twenty-first century.

As Sreekumar, in Chapter 4, rightly argues, the grand narratives of development have been replaced with seemingly fragmented, but powerful, discourses of participation, empowerment, micro-enterprises, self-help, good governance and a host of other ambiguous interventions and concepts defined and practised by an emerging nexus of international development bureaucrats and civil society professionals across the developing world. What the meta-models of development failed to deliver with industrial technologies, ICTs and other new technologies are expected to deliver for the cyber-libertarian micro-development initiatives across the developing world. The community-based, self-help micro-enterprises are deemed to deliver if only states and international donor agencies could bridge the digital divide and knowledge gap that persist within developing countries and between developing and developed countries. Therefore, the default solution for the digital divide is supposedly the widespread diffusion of ICTs in the developing world to enhance connectivity to the technological wherewithal and markets of a rapidly globalising world.

However, the exploitation of ICTs in tandem with economic globalisation that heralded information capitalism has spurred renewed questions about equity in social and economic development in countries like India. For example, the economic growth encouraged by IT and other high-tech industries is being confined to the urban areas in parts of the country. India is 'shining' only for the urban middle class in cities like Bangalore, Hyderabad, Chennai and Mumbai (contrary to the claims of the previous BJP government that was voted out of office by the disaffected rural poor in India), while the majority of the Indian people living in urban slums and rural areas did not benefit from this growth. It appears that information capitalism is a far more complex and uneven process of economic change and development than industrial capitalism because of the complex global–local forces and factors that underpin its dynamics.

Digital divide, unequal growth patterns, increasing income gap between urban and rural areas and other issues born of the distributional consequences of information capitalism must not be seen as a mere access (to cyberspace) issue, but as an equity issue. For example, those who seek solutions to the digital divide see the problem as merely an iteration of the logic of economic growth and expect to find the solution through providing the means (technology). The problem must be seen as a rupture in the iteration of modern capitalism, which is what students of the political economy of information capitalism must address to square theory with grounded realities. While the political economy question of digital capitalism aids comprehension of the problem of digital divide and related equity issues, the social and cultural context within which the technology diffusion takes place is also key to understanding the problem and finding meaningful solutions.

Pertaining to the central question of the prospects of rapid economic development consequent to the development of certain ICT industries in India, we may ask how optimistic students of the political economy of Indian development are about the durability of this new development paradigm. Can we treat economic development due to the export of software services and ITES as a new mode of development? Why is IT so mesmerising at the surface but pregnant with a host of contradictions? Does the information technology-driven so-called 'new economy' bypass the domestic economy (especially the large component of it affecting the rural masses) and is it inherently global in origin? Is information capitalism radically different from mass and bulk production (manufacturing)-based industrial capitalism? What sort of international division of labour exists in informational capitalism? What sort of flexibility and social mobility exists in the IT labour market? What is the true extent of India's capabilities in the software services and ITES? Is India poised to leapfrog into the Information Revolution without going through a necessary Industrial Revolution?

Chapters in this volume will address these theoretical and empirical issues and concerns. Since India is touted as a 'shining' example of a major developing country that has apparently benefited from IT-driven economic growth, a collected volume on Indian experience with informational capitalism is highly topical and timely. Lessons from India's experience would be extremely useful for other countries that are trying to emulate India's IT policies to replicate its experience. While some chapters address the question of knowledge and digital divides and their consequences in a historical as well as the contemporary context of the evolution of digital capitalism, other chapters address such issues as the viability of IT-enabled outsourcing as a development paradigm; state–civil society divide in community informatics in India; labour control of immigrant Indian workers in the IT industry; competitive political implications of the rural–urban digital divide; political economy of IT investments and development strategies, and comparative competitive strategy in IT development and its sub-optimal innovation trajectories.

In Chapter 2, Anthony D'Costa analyses the recent growth of the Indian ICT industry, particularly its export dynamism, and raises a host of questions pertaining to the durability of this new mode of economic development. For example, D'Costa asks, is export-led growth in an increasingly integrated world economy the best way to usher in economic change? What is the significance of technical education and skill development for technology upgrading? Can transnational corporations catapult the Indian industry to global competitiveness? Is the state limited to infrastructure provision within a strictly market-augmenting role? Is the Indian industry sufficiently innovative to remain internationally competitive? Is IT a 'carrier technology' that can usher in a large-scale transformation of the Indian economy? These are relevant questions and deserve close scrutiny.

What is curious however, is an accompanying sense of euphoria within industry, government, and even academia, which views the Indian software industry as a panacea for India's developmental problems. Correspondingly, supporters of the industry suggest that India's unprecedented growth in software exports represents a new model of economic development, which, if leveraged correctly, will go a long way to lift the Indian economy from its infamous 'low level equilibrium trap' to a new trajectory of sustained growth and structural change. There is no doubt that the Indian software industry is dynamic and presents an unprecedented opportunity for economic change. As economic development is a process of increasing value addition by using higher skills, it is natural that software industry expansion in India will evoke favourable responses. Similarly, economic development entailing structural change – from agriculture to industry to services – suggests the leapfrogging nature of Indian development trajectory. Clearly certain regions, social groups and economic sectors have benefited from this sector. However, mesmerised by the industry's rapid expansion, the magnitude of the developmental challenges in India has been underestimated, according to D'Costa, while the industry's transformative capability has been overestimated.

In Chapter 3, C. P. Chandrasekhar points out that even before the sustainability of the rapid expansion of the software industry could be established, albeit from a low base and at the expense of the hardware sector, the focus of attention in the broad area of the political economy of IT and development has shifted to the outsourcing phenomenon.

According to Chandrasekhar, outsourcing is being presented as the emerging development paradigm for the most dynamic segments of the organised services sector of the so-called new economy. As has been true of other areas of the IT sector, the hype over outsourcing has resulted in two tendencies. The first is reflected in the neglect of the fact that outsourcing is an old and heterogeneous phenomenon. It has occurred before in manufacturing; it could be of software generation or of ITES that are poles apart in terms of activity attributes; it does not dis-tinguish between 'outsourcing' and 'outlocation'. Each of these has substantially different implications for the development impact and sustainability of the phenomenon. The ambiguity is the greater because the same or related firms are involved in multiple activities. The second is reflected in the fact that the rapid development of outsourcing has resulted in significantly divergent and completely unjustified figures of the actual and potential size of revenue and employment. This, when combined with ambiguity, results in gross over-estimation of the size and structure of the software sector. Consequently, it becomes extremely difficult to assess the actual development impact and sustainability of the IT sector boom in an economy like India's. Chandrasekhar conceptually disentangles the outsourcing/outlocation phenomenon, critically assesses the figures being put out by various industry protagonists, ponders the pattern of evolution of the phenomenon and, finally, assesses the developmental impact of alternative paths of economic transformation.

Drawing on insights from the recent development debates within the realm of community informatics, in Chapter 4 T. T. Sreekumar seeks to analyse the political economy of ICT-based development projects initiated by civil society organisations (CSOs) in rural India. His objectives are to analyse three interrelated aspects that require closer scrutiny in the context of the rise and prominence of ICT-based rural initiatives in India, namely, the mediation of technology and power in ICT-based projects, their development claims and achievements, and proclaimed promises of e-governance. The mediation of technology in the context of rural ICT initiatives has to be understood: (1) within the context of the political economy of digital divide; and, (2) within the domain of empowerment mediated through and negotiated by local politics. A closer look at the claims and achievements of ICT-based development projects as initiatives for enhancing capabilities and forging successful alliances with governmental efforts to redefine rural interventions for community participation, empowerment and self-help projects provides the backdrop for the analysis. The chapter, based on

Sreekumar's fieldwork at several 'TARAkendra' information kiosks in Bhatinda, Punjab and the Information Village Research Project kiosks in Pondicherry, unravels a cyber-libertarian turn in development studies, which reflects a convergence of development rhetoric and information society theories in discourses on the digital divide.

In Chapter 5, K. J. Joseph shows the perils of excessive export-orientation in India's recent development strides within the context of the export boom in the software services and ITES sectors. Drawing on insights from 'Dutch disease economics' in trade theory, Joseph highlights the opportunity cost of channelling valuable resources to induce this export boom on other sectors of the economy. The macroeconomic implications of IT export boom, in a more precise way, may be viewed from two angles: implications on the product market and implications on the labour market. The central argument of the Dutch disease model is that the income windfall from an export boom can cause domestic economic problems. It can even lead to the most unexpected outcome of de-industrialisation of the economy. The boom in the software sector, needless to say, would have led to an increase in the demand for labour on the one hand and increase in the wage rate on the other. The increased demand for software professionals, in turn, would induce the professionals with specialisation in areas other than software to migrate to the software sector. Joseph argues that the export-oriented strategy, undoubtedly, has made India a major exporter of software and ITES. The perils of such a strategy are that countries using IT tend to benefit more than IT-producing/exporting countries. Therefore, to mitigate the adverse effects, large-scale diffusion of ICTs to other sectors of the economy is essential. That calls for an urgent national policy on ICTs. On the whole, it appears that while India is a major ITES and software services provider, its progress in overall ICT use has been remarkably lacklustre. Joseph claims that Indian policy-makers have failed to realise that the social marginal product of a dollar's worth of ICT used at home is bound to be much higher than a dollar's worth of ICT exported.

In Chapter 6, Jayan Jose Thomas looks at the various factors associated with the diffusion of ICTs in rural areas, and in doing so he looks at the potential role that ICTs can play in the development of rural areas. Thomas argues that on the one side we see the phenomenal growth of the ICT industry in some Indian cities, and on the other side we see the continuing impoverishment of India's rural poor. The case of the Indian state of Andhra Pradesh is particularly telling in this regard. Andhra Pradesh, and in particular its capital city Hyderabad, received widespread media attention in recent years for attracting sizeable investments in the IT industry. Over the same period of time Andhra Pradesh has been passing through a phase of unprecedented rural distress, including cases of starvation deaths, distress migration of landless agricultural labourers and suicides among farmers and handloom textile workers. Thomas's major conclusion is that ICTs can play a potent role in rural development, but only if the basic obstacles to rural prosperity are removed through radical changes – through land reforms, rural credit, and greater state intervention in rural infrastructure and primary education. Empirical support for the chapter is based on field studies conducted in July/August 2004 in two rural locations in two South Indian states – Kuppam in Andhra Pradesh and Malappuram in Kerala.

Continuing the focus on Andhra Pradesh, in Chapter 7 Kyle Eischen argues that this Indian state is an important site to understand the impact of information capitalism in India. Eischen shows that a combination of global capital and labour networks, local entrepreneurs and regional government initiatives has propelled the state from virtual non-existence to an increasingly central node of software and IT services in the global economy. These patterns – building upon past national legacies and an expanding global industry – are indicative of both the broader changes within India during the 1990s and the potential for information industries to thrive in unexpected and new environments. Yet, in spite of success within the global industry, the 2004 elections witnessed the electorate's resoundingly removing from elected office the chief architects of the regional policy in Andhra Pradesh – the Telugu Desam Party led by Chief Minister Chandrababu Naidu. Equally important, the electoral change extended across India to both national and state government leaders that had presided over and emphasised India's emergence as a key actor within the global IT industry.

This contradiction – success in building an informational industry yet failure with the electorate – suggests it is important to re-examine the nature and impact of informational development in India generally and Andhra Pradesh particularly. Eischen addresses this issue through two interconnected aspects surrounding growth during the 1990s. First, while the IT industry has grown, it still remains unclear in most analyses exactly what spurred such growth. Simply success for specific firms or even the regional industry does not necessarily signal widespread and general economic development. Second, there were significant market, policy and industry shifts during the 1990s that need to be correlated with on-the-ground transformations. Shifts in policy and industry formation, in particular the emergence of ITES during mid-1990s to 2005, signal the limitations of earlier models and a move towards new solutions by both policy-makers and firms. That such shifts, argues Eischen, still did not significantly alter the outcome of the recent elections presents an opportunity to understand the possibilities of informational development more generally.

In Chapter 8, Balaji Parthasarathy explains the opportunities and constraints created by the increasing globalisation of the Indian software industry and the organisational changes that have accompanied the growth in software exports since the mid-1980s. Technological change since the 1970s until the time of writing has sustained the global demand for software that advanced industrial countries have been unable to meet. This created opportunities for new producers and players such as India thanks to the availability of a relatively large, low-wage high-skilled, English-speaking labour pool, and increasingly liberal domestic economic policies. After entering the international division of labour by providing low-wage, low-skill onsite services, the Indian industry began increasingly to shift more value-added offshore services in the 1990s to become the largest non-OECD software producer by 2000. Post-2001, the industry continued to grow despite a global slump in demand by capitalising, in part, on the globalisation of R&D services. Despite the transformation in the organisation and the capabilities of the industry, however, the export of profitable software products and packages is limited. Besides explaining the differential success in services and products, Parthasarathy discusses the relevance of the Indian case for other newly industrialising countries that are trying to board the global software industry bandwagon.

While globalisation has freed capital to move around in real time to seek profit wherever it can, little attention has been paid to the labour that sustains this industry. While existing literature has, more or less, grasped important features of the current world economy, particularly its volatility, and has gone quite far in identifying the causes and consequences of the increasing uncertainty at the macro level, little has been done to tackle how the volatility itself is socially constructed through concrete human activities. In Chapter 9, Biao Xiang takes a macro-sociological approach to explain the contingencies thrown up by the uncertainty associated with how labour copes with the globalisation of capital. Labour is essential for the IT industry not only because the industry is labour-intensive, but also in the sense that creating a labour system responsive to the extremely volatile capital movement, hardly experienced by other industries, constitutes the most challenging task for the IT industry as a social process.

Based on his fieldwork among migrant Indian IT workers, Xiang demonstrates that central to labour management in the IT industry, at least in the present case, are three processes. First, there is a process whereby IT companies outsource labour management tasks to a placement agency, which enables the IT companies to respond to the volatile financial market and the ever-changing technology trends as flexibly as possible. The second is the process of interaction between placement agents and workers through which the former exercise direct control over labour on a day-to-day basis. Finally, there is a process whereby individual workers move up a career ladder within the existing institutional system, typically by acting as sub-agents of other placement agents or setting up their own placement agency, which in turn enhances the labour control mechanisms.

Finally, in Chapter 10, Govindan Parayil reflects on the social and economic inequalities of informational capitalism by examining two contradictions of ICTs-led economic development – increasing returns and digital divide. Two main and interrelated strands of evidence are presented: first, contrary to expectations that rising income per capita will tend to reduce wealth and wage disparities, the distribution of income and wealth both between countries and individuals has increased in the information age; second, knowledge production is a self-reinforcing cycle that tends disproportionately to reward some and to exclude others. The so-called digital divide is as much a symptom as a cause of these broader techno-economic phenomena, and regarding it as a simple issue of connectivity is simplistic and reductive.

All the chapters in this volume are at the time of writing timely reflections on the political economy of information capitalism in India. All chapters reflect on the equity implications of information capitalism. It is expected that the debates initiated on these pages will continue through debates and policy interventions at various levels of industry and government to bring the immense potential benefits of ICTs to all Indians, especially the rural poor.

2
ICTs and Decoupled Development: Theories, Trajectories and Transitions

Anthony P. D'Costa

1. Introduction

The objective of this chapter is to demonstrate that the trajectory of recent Indian development under the spell of information and communications technologies (ICTs) reveals a number of debilitating structural flaws caused by a host of external and internal factors. The trajectory of the Indian ICTs sector is unique, both cross-nationally and relative to its own historical record. Few countries have been able to match the growth rate of the Indian ICTs sector. However, outside this sector few high-skill industries in India display international competitiveness. In the larger economic system there appears to be no industrial upgrading that parallels the software export boom. In other words, the industry is not associated with the economy-wide technical change and structural transformation that was experienced by other late industrialising countries. Instead, the Indian IT industry seems to have settled comfortably into capturing a good share of a particular global niche, namely software services (D'Costa, 2003b).

Conceptually, this could be argued to be an 'enclave' form of development, not all that different in a structural sense from colonial exports of the past or the more recent exports of manufactures from export-processing zones. There are, of course, differences. Software, though labour-intensive, is a high-wage, high-skill activity. Advanced science and technology education is a prerequisite for its development. The industry has considerable links to related convergence technologies and high-technology manufacturing industries. Yet it can be shown that this enclave form is a result of the cumulative interaction of the global and the Indian national economy, which is facilitating an expansionary trajectory, but not one that is wide enough to include other sectors, most geographic regions or most social groups. This is a classic problem of uneven and combined development (D'Costa, 2003c). More narrowly, the enclave nature of the industry could manifest itself in several specific forms of decoupling, as reflected in the sector's weak links to a domestic market and the absence of a vibrant hardware manufacturing segment. What this suggests is that the industry's enclave form is likely to limit its larger transformative capability.

The narrow success of the Indian ICT industry is the outcome of three interrelated processes, namely, decoupling, path-dependence and lock-in. Decoupling refers to the software industry's being weakly linked to domestic markets (other manufacturing and service sectors) or, conversely, strongly coupled with foreign markets. Either implies limited spin-offs to induce wider economic transformation within India. The concept of path-dependence is also well established. It refers to the importance of history, mainly institutional history, in shaping contemporary dynamics, suggesting that events in the past can have considerable impacts down the road (Rosenberg, 1994; Arthur, 1994; Antonelli, 1999). For example, India's long-standing education policy biased towards tertiary technical education for industrialisation has contributed to the large supply stream of technical professionals, which in turn facilitated the emergence of the ICT industry.

Relatedly, lock-in effects have to do with a particular set of economic and sector-specific dynamics that is self-reinforcing. For example, India's strong outsourcing arrangements with United States firms, while commercially lucrative, have constrained the innovative capability of India's own software development (D'Costa, 2002a). India's high share of exports to the US, though a small percentage of the total US ICT market, could have a large political impact at a time when well-paid high-tech jobs in the US are becoming scarce, taking the form of restrictions on trade and outsourcing imposed by the US. This could adversely affect Indian firms that are not catering for diverse geographical and product/service markets in the short term, and undermine India's learning experience in the longer term. At the same time, the enclave nature of India's success in the IT industry could be a harbinger of pronounced sectoral inequality between high-tech services and the rest of the economy, as well as uneven regional development within the country (D'Costa, 2003c).

Section 2 of this chapter presents some principal characteristics of the Indian ICT industry. Section 3 links the development trajectory of the Indian ICT industry to some received theories of economic development and their practices, focusing particularly on structural change.

References are made to the East Asian cases to illustrate the significance of inter-industry linkages for economy-wide transformation. The final section broadly suggests how ICT could be made a new engine of growth and structural change, by bringing to the fore more traditional development concerns such as income distribution, mass education and infrastructure development.

2. Key features of the Indian ICT industry

Global competition suggests that not all firms can or should undertake all activities. Rather, outsourcing, joint ventures, subcontracting, technical collaborations and alliances are institutional arrangements that firms adopt to specialise (Okhi, 2001). In practice, large multinational corporations (MNCs) outsource or subcontract some of their non-critical production and service needs to other firms. This presents two opportunities, as new markets are created. Indian software firms have leveraged these opportunities. Many Indian firms provide software services to Fortune 500 companies. However, there is a downside to this type of economic participation. By forging strong links to external markets, local firms depend on foreign firms and external markets for their growth. While this is not an unfavourable development *per se*, in the absence of dynamic domestic demand the anticipated spillover from export activity into the domestic economy can be severely truncated. This decoupling of the Indian software industry with the local economy manifests itself in four ways:

1 The industry's growth is not based on domestic demand, though the technical talent is a product of the national economy. It is heavily export-oriented: nearly 60 per cent of India's IT output is exported. Roughly two-thirds of India's software and service exports are aimed at the US market. The growth rate of the domestic IT market is well below the growth of the export market.
2 The industry is not linked to a domestic hardware manufacturing sector, a segment that typically generates considerable spin-offs. While there are commercial and technological reasons for a weak hardware segment, the net effect is a heavy drain of technical talent from other manufacturing sectors.
3 The industry's specialising in software services, which is beneficial if raw employment of native talent is concerned. However, that market segment is the least dynamic in terms of future growth (NASSCOM, 2002).

4 As not all export markets are the same, the Indian industry, by locking itself into that of the US, foregoes any learning opportunities that would come with engaging other geographic markets. Japan is a case in point despite the many institutional and policy barriers that come bundled with the Japanese market (D'Costa, 2004b).

The decoupling of the industry can be partly explained by industry-specific dynamics occurring outside India. Recent advances in ICTs show a unique trend in that software production ceases to be linked with hardware production. Microsoft, as the leading monopoly player in packaged software, has set the standards for all Windows-based software applications. Microsoft's Windows operating system along with Intel's microprocessor chips have become part of a dominant design characterised as 'Wintelism' (Hart and Kim, 2002). Consequently, other hardware-independent application software producers are compelled to develop Windows-based applications (Hart and Kim, 2002). Software, however, is complementary to hardware. The latter includes electronic data-processing equipment (computers), office equipment, controls instruments, communications equipment (radio, TV, mobile systems), consumer electronics and components that go into such hardware. India's position in the global ICT market is low (see Table 2.1). While this is consistent with the 'loosely coupled technological system' associated with the ICT industry (Hart and Kim, 2002, p. 3), the lack of a domestic production base generates a different kind of industry trajectory and, possibly, induces distortions in the larger economy.

India has essentially missed this ICT-related manufacturing bus. Barboza (2004) presents an interesting graphical illustration of the various countries involved in making the components for an IBM PC. These include Thailand, Malaysia, Taiwan, Korea, Mexico (for assembly), the US and Canada. India is conspicuously absent despite its global presence in the software industry. East and South-East Asian countries dominate the consumer electronics and computer production chains. In this global division of labour, India ranks only 26th among 33 countries. Indian software producers in the early 1990s believed that by the end of the decade India's hardware industry would be comparable to Taiwan's (Lakha, 1999). In 1999 Taiwan's ICT goods production was ten times that of India's. Smaller countries such as Ireland, the Philippines, Thailand and Israel had greater production in these segments than India. India's import substitution industrialisation had protected an incipient hardware sector in India during the 1970s and 1980s. However, the economic reforms of the 1990s exposed India's hardware

Table 2.1 World production of ICT goods (selected countries)

	1999 total ICT ($mill.)	(%) share of world total	1990–99 (CAGR*%)	1995–99 (CAGR %)	1997–99 (% change)
US	320,840	29.5	5.8	4.4	20.3
Japan	220,728	20.3	2.3	4.0	1.3
S. Korea	57,597	5.3	10.8	4.2	19.2
Germany	47,545	4.4	0.5	4.4	37.9
Ireland	16,481	1.5	12.8	17.0	54.9
UK	47,734	4.4	6.2	6.9	28.4
OECD-21	843,121	77.5	4.2	1.4	16.5
China	59,738	5.5	19.8	20.1	NA
Malaysia	38,956	3.5	20.1	9.0	32.8
Singapore	40,755	3.7	11.9	0.7	–4.7
Taiwan	40,979	3.8	12.7	9.0	30.6
India	4,841	0.4	0.4	–0.3	5.5
Brazil	13,484	1.2	1.3	–5.0	–30.4
Israel	6,311	0.6	14.6	11.2	35.5
Total	1,088,539	100.0	5.6	1.4	22.3

Source: Adapted from OECD, 2002, 2000a
Notes: CAGR = compounded annual growth rate

sector to strong international competition, undercutting its technological capability (Sridharan, 1996).

In contrast to the hardware sector, India's software sector presents a contrasting picture (see Figure 2.1). India's ICT industry is spearheaded by the rapid expansion of software for exports. In 1995 the industry had under $2 billion in output. In 2003/2004 it had reached nearly $20 billion, representing 3.82 per cent of India's GDP (NASSCOM, 2004). During 1994–2004, India's exports of software and software services increased from less than half a billion dollars to $12.2 billion, representing over 21 per cent of India's total exports in 2003/2004. What is remarkable is the steady decoupling of India's hardware from software production, evident from a rising ratio of software exports relative to hardware output (D'Costa, 2003b).[1] Indian software production is also decoupled from the domestic market (Kattuman and Iyer, 2001). India also has an unusually high share of software production relative to the total IT market in India, ranging from nearly 50 per cent in 1995 to over 75 per cent in 2002.

The third form of decoupling is India's specialisation in software services, an activity quite distinct from developing software packages. Soft-

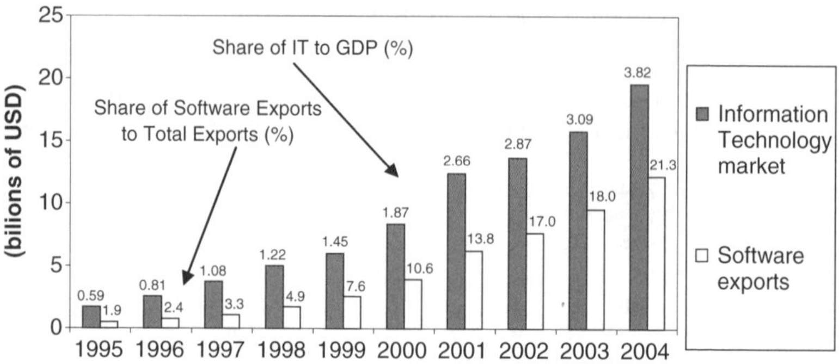

Source: NASSCOM, 2002, 2004. Notes: IT = hardware, peripherals, networking, domestic and export market for software and services, and IT-enabled services.

Figure 2.1 Recent growth of the Indian IT industry (1995–2004)

ware packages are more profitable as they provide a steady stream of revenues once the basic product has been successfully launched. Software services, on the other hand, tend to be stand-alone projects. A thriving IT industry ought to have a good balance between the two types of software production. The spectacular growth of Indian exports, averaging over 50 per cent in most years, has been through software services. These exports reached nearly 75 per cent of the total Indian IT market in 2001/2002 and have been estimated to be 62 per cent in 2003/2004 (NASSCOM, 2004). This relative decline is attributed to changing definitions of IT market niches accompanying the disaggregation of output data. What is missing from the basket of software exports is software products or packages. This is in sharp contrast to nearly $4 billion of exports of packaged software by Ireland.[2] There is no comparable figure for India, suggesting a negligible presence in this market niche. The closest India comes to products is installing and integrating 'packaged software' created by multinational firms. Such exports from India were estimated to be worth only about $350 million (NASSCOM, 2002).

The fourth form of decoupling is illustrated by India's exclusion from the Japanese market. While Japan's market size relative to the US and the European Union is small, its absolute size is nearly $35 billion. However, India's share of this market is about one-half per cent compared to India's shares of nearly 4 per cent and 2 per cent of the North American and European markets respectively. More importantly, Japan's dependency ratio is very low, suggesting a variety of structural and institutional factors involved in excluding Indian exports to Japan. What this means is a less diversified learning environment for the Indian industry (D'Costa, 2004b).

India's inability to penetrate other market niches such as software packages is due to several factors. First, product development requires a dynamic domestic market. Second, without a large installed base of hardware at home the benefits of network externalities cannot be captured, thus limiting the market size and future product development. Third, penetrating the foreign software package market is daunting due to development and marketing costs. Packaged software development demands intensive interfacing between users and producers, which neither small local markets nor geographically distant consumers can support. Fourth, software products are subject to first-mover advantages with dominant design effects. Most small firms are at a disadvantage; developing country firms are further disadvantaged. Hence it is not surprising to find that the bulk of India's software exports are in services and a good part of those are in low-value subcategories of legacy management, customised services and IT-enabled services (ITES) (NASSCOM, 2002).[3] According to NASSCOM (2002), the global demand for IT services, which includes customised services and ITES, will be over $700 billion in 2005. India's current strength is in customised services, which is likely to be sustained in the future. However, this segment of the global IT services market has been estimated at only 4.4 per cent of the global market in 2001 and is expected to fall to 3.6 per cent by 2005 (D'Costa, 2003b).

If the market segment in which India excels is small to begin with and is likely to witness a relative decline in a few years, it is clear that India needs to diversify its markets away from this segment. This may mean moving into high-value software services for which there are considerable opportunities, such as systems integration, packaged software support and installation and processing services. But it does not mean that India should give up low-end services. For example, low-wage ITES is a growth sector with significant absorptive capacity for India's large pool of educated, non-technical personnel. In which direction India's information and communications technology (ICT) industry moves would depend on the supply of skills in India, competition from other low-wage economies and their global demand (Desai, 2001). What is incontrovertible, however, is that India must not only develop higher-order skills in general but also widen and deepen its experience in complex project design and execution so as to cope with the rapidly changing global ICT market.

3. Economic transformation in theory and practice

Given the background of the Indian software industry, the industry's positioning can be interpreted as a consequence of a particular business model (as distinct from, but not necessarily in conflict with, a development model). As with the rise of new ICT-related opportunities (quite independent of India's technological capability and home market growth), the availability of a particular kind of (English-speaking) skilled worker in India allowed US firms to outsource their software needs. It began with onsite services (known as 'body-shopping', see Chapter 9), whereby Indian professionals worked at the client's base. These projects, though highly profitable, were low-revenue and tedious and they tended to under-utilise the technical talent (D'Costa, 2002b). The business model gradually shifted towards offshore development, whereby, increasingly, software projects were carried out in India itself on behalf of foreign clients (D'Costa, 2002c). This was more lucrative, as costs were lower in India.

This model emerged during the era of neo-liberal reforms, which encouraged global integration. Initially low labour costs played a very important role. They still do. However, gradual skill upgrading contributed to India's larger global presence. Increasing returns have set in as more global ICT firms are outsourcing software development, some design activities and business processing operations (BPOs) from India. Recent economic downturns have encouraged global outsourcing even more, as foreign clients are compelled to slash costs further to maintain their profit rates. Fear of job losses has prompted some states in the US to entertain the possibility of restrictions on outsourcing by US firms (Rai, 2003). This raises the old question of whether exports can be an engine of growth in India and, more crucially, whether a single export-oriented subsector can pull the rest of the Indian economy towards meaningful structural change.

From an economic perspective, development is concerned with macroeconomic structural dynamics, whereby there is a progressive shift away from lower value economic activities towards those technologically more complex. Hence, development occurs when there are broad, durable movements from agriculture to industry and services, and within industry from labour-intensive to more capital-, technology- and skill-intensive industries. Amsden (2001) refers to this as the process of progressively creating more knowledge-based assets. In this process of structural transformation it is evident that both market forces and institutional interventions facilitate the process. The neoclassical interpretation is that there are stages by which particular broad sectors will emerge with the aid of price signals best generated by free markets. In this account, the rising cost of labour in agriculture due to increasing demand for such labour is likely to allocate labour in more profitable industrial sectors. Structural rigidities such as resource immobility are assumed away. A structuralist account *à la* Arthur Lewis would suggest strategic intervention in industrial sectors to overcome the debilitating effects of low productivity in agriculture, effectively skipping (or compressing) some stages of development. Many others have argued in favour of state regulation to alter prices deliberately and encourage rapid industrialisation and technological upgrading (Amsden, 1989; Moreira, 1995). Experience from several late-industrialising countries suggests that relative labour scarcities that arise due to an increased rate of labour-absorbing activities do push the economy to higher value-added activities. The experience of East Asian late industrialising countries demonstrates how state intervention and firm initiatives can produce the desired results.

The transformative task of the state goes beyond nurturing internationally competitive firms to effectively managing the transition from labour-intensive industries to more capital- and technology-based sectors. Three features of this development process are noteworthy: broad-based development; effective state promotion of key sectors; and channelling competitive pressure from export markets to national innovative capability. East Asian economies offer some lessons on how this can be achieved. They regulated markets without undermining robust capital accumulation; ensured better asset and income redistribution through land reforms and labour-intensive industrialisation; and sought integration with the world economy for accessing technology. For example, rising productivity and the pressure on wages compelled East Asian firms to upgrade technologically and diversify away from low-value, labour-intensive goods. In this transformation the domestic market played an important role, particularly for the establishment of heavy industries and the subsequent transition to knowledge-intensive industries.

This cost-push, interventionist model for industrial upgrading in East Asia is now well established. There have been leading 'clusters' of related industries that induced resource-scarce economies to move into labour-intensive export-oriented manufactures, and, subsequently, capital- and technology-intensive production for both domestic and foreign markets. The promotion of heavy industries such as steel, petrochemicals, automobiles, ship-building and so on is one such example of an industrial cluster in both Taiwan (Big 10 Projects) and South Korea (Heavy and Chemical Industries Programme). Taiwan, given its particular historical and institutional parameters, relied more on small firms to spearhead the high technology industry. There was considerable local effort in learning while drawing on the huge inventory of global sources of innovation.

The East Asian economies compressed the development process by speeding up growth through productivity growth, industrial diversification and upgrading, and technology learning and innovations. The state was instrumental in incubating the entire range of manufacturing industries from the simpler labour-intensive to the more complex skill-based high-technology. They have also developed a relatively productive agricultural sector. The East Asian economies moved away from the neoclassical concern of efficient allocation of a society's scarce resources. They followed the Ricardian division of labour based on comparative advantage (lower relative costs of production brought about by efficient allocation of a society's resources) only in a limited sense. Instead, careful orchestration of industrial and technological upgrading through logical sequencing of industrial strategies created the basis for transition to a new economic structure.

The Indian development model also included state intervention, aimed to compress the stages of industrial development. However, due to its particular political economy, the state did not aggressively pursue a performance-based reward system, effectively making Indian industries uncompetitive (Chibber, 2003). But it was not just state failure that was the problem; the market failed too. The high incidence of poverty, income inequality and illiteracy are serious constraints to using a knowledge-intensive sector to induce economy-wide transformation, unless backward and forward integration to inputs and users are strong and growing. India's poor record in poverty alleviation and education means that only relatively affluent families with well-educated household members will benefit from the ICT industry. While the incentive for tertiary education is high in India, the market does not work for the poor. The tight links of the industry to universities and technical institutes in India are very selective and do not include the masses. Nor are there strong links to related industries in the ICT sector such as manufacturing of computers, other hardware, electronic components, monitors and LCD screens. Similarly, from the users' side, the high cost of computers and software restricts purchases to the better-off households and businesses. Clearly, from a policy angle focusing solely on software exports and only customised software services is not adequate to transform the Indian economy.

The business model (offshore development centres) adopted by the Indian ICT industry is consistent with globalisation and economic integration. It is similar to the outsourcing arrangements of the East Asian economies. The difference, however, is that the software clusters are tied heavily to the foreign market on the user side and only narrowly integrated backward to other sectors. The absence of a hardware industry as well as a large user base limits revenue as well as technological spin-offs. In both Taiwan and Korea high technology manufacturing includes the entire gamut of hardware and software, with considerable backward integration (Dedrik and Kraemer, 1998). It is not just the assembly of PCs or notebooks but also the manufacture of memory chips and LCD panels and the writing of embedded software that constitute the cluster. The Taiwanese experience is illustrative of the sequencing of industrial upgrading within the electronics industry, not only making use of its US-trained techno-entrepreneurs but also encouraging them to establish start-ups in science and technology parks. Such industrial sequencing within the broader ICT industry suggests a purposeful build-up of technological capability and the institutionalised nature of the learning process.

The link between the software segment and tertiary and technical educational institutions is strong in India. Just as East Asian governments have been critical in industrial upgrading, the Indian state, unwittingly, has been behind the rise of the software sector. The offshore development model would not have been possible without state-provided infrastructure support for software technology parks. In today's globalised economy, exports are no doubt an important driver of economic expansion. The Indian industry is contributing significantly to this effort. But the Indian ICT industry cannot be perceived to represent development leapfrogging, that is, it does not negate the significance of routine aspects of development needs such as income, education and health, which expanding agriculture and more traditional industries could foster. The vitality of any economy resides in the increasing complexity of economic activities, the associated dense intersectoral linkages mainly at home, and their articulation with the global economy. The Indian software sector fulfils the last condition quite well, but only weakly meets the first two. What the state has done is provide a favourable policy and infrastructural environment for Indian firms to export software services and allow multinationals to outsource from India. What the state needs to do is work towards promoting intersectoral linkages by widening and deepening the domestic market for ICT and non-ICT goods and services. Social spending in the areas of education, health and infrastructure to meet the millennium development goals combined with a sound technology policy meeting new market challenges in the ICT industry will be critical for a more robust development experience driven by the ICT sector.

In India's ICT sector, the software industry is driven by external demand, and hence labour costs in this sector will tend to be higher than in other sectors. Similarly, the cost of labour is likely to rise much faster than costs in other sectors, raising the possibility of internal brain drain from other sectors. There are economic costs involved in this resource reallocation mechanism, especially if high-skilled labour more suited to non-ICT sectors is involved. It is possible that the resulting relative scarcity in non-ICT sectors could compel firms to become more productive, thereby generating overall technical change. However, the supply of talent is not elastic in the short to medium term. Furthermore, non-ICT sectors could be sufficiently weakened by competition and skill shortages to make technological resurgence difficult. Hence, it will be necessary to have a more proactive strategy of ploughing back high export revenues earned by the software industry to innovative activities, such as importing foreign technologies, adapting them to local conditions and needs and developing the infrastructure to assimilate them.

Given the structure of the Indian economy, it will be decades before ICT can become the engine of technical change for structural transformation (see Table 2.2). India remains largely agrarian. Its per capita income is one of the lowest in the world, with its purchasing power parity income somewhat better than nominal per capita income expressed in US dollars. Income distribution, which is better than in many other developing countries, is getting worse. International competitiveness, as revealed by export profiles, suggests India's continued dependence on low-wage, commodity-based exports.

The significance of ICTs, however, should not be underestimated. As noted earlier, an important item in India's export basket is software services. The share of IT and software in India's GDP and exports has been rising. In 1995, the IT share was 0.59 per cent of GDP, which in 2003/2004 increased to 3.82 per cent. Similarly, from 1995 to 2002, the share of software exports to total exports rose from 1.9 per cent to 21.3 per cent (NASSCOM, 2002, 2004). This is evidence of limited structural change as more of the economy's output is shifting towards IT, which is a composite subsector of manufacturing and services. Is this the beginning of a long-term trend? Does this mean the Indian ICT sector is poised to fundamentally shake up the Indian economy?

Table 2.2 Structural changes in the Indian economy, sectorial contribution to GDP (%) 1950–2004 (1993/94 prices)

	Agriculture	Manufacturing	Transportation	Finance	Public Administration
1950–51	59.19	13.29	11.95	6.69	9.41
1960–61	54.75	16.61	13.74	6.11	9.17
1970–71	48.12	19.91	15.55	5.94	10.69
1980–81	41.82	21.59	18.41	6.53	11.65
1990–91	34.92	24.49	18.73	9.69	12.18
2000–01	26.25	24.90	22.80	12.59	13.46
2003–04*	24.43	24.58	24.75	12.83	13.41

Notes: GDP at factor cost by industry of origin. *Estimate. Agriculture includes forestry, logging, fishing, mining and quarrying. Manufacturing includes construction, electricity, gas and water supply. Transportation includes trade, storage and communication. Finance includes insurance, real estate and business services. Public administration includes defence and other services
Source: Government of India, Ministry of Finance, 2004, S-5

The world demand for software services and ICT goods can be assumed to grow. The question is whether the Indian market will be a significant driver of the expansion of the ICT industry from the demand side, which, in turn, will become a driver of economy-wide technical change. At the industry level, technical change produces knowledge-based assets. In a modern economy firms generate such assets mainly through manufacturing and high value-added business services. It is difficult to model the links of the microeconomic processes of business innovations to the macroeconomic impact of such innovations (Beije, 1998). However, technical change means that most economic and industrial sectors absorb new innovations and become more efficient. It implies dense, self-reinforcing intersectoral linkages, fuelled by the diffusion of ICT-producing and ICT-using innovations among the various sectors. Measuring the efficiency gains from the application of such innovations is difficult. However, we know from some of the OECD economies that ICTs have a favourable impact on productivity growth (OECD, 2000b). We have also established that the Indian ICT sector is weak on hardware manufacturing. And the best performing segment of ICT, the software sector, is largely export-driven. Hence, in the absence of a dynamic ICT manufacturing sector and local demand, the software industry's contribution to India's knowledge-based assets will continue to be constrained.

Another way of viewing the problem is the lack of real connection between production and consumption. A recent OECD report stated that countries could reap benefits from strong ICT-producing and ICT-using sectors of the economy (OECD, 2002). The Indian industry does well on the former but not on the latter, suggesting that exports will lead to domestic market development rather than the other way round. Theoretically, this is possible only in the initial rounds. In subsequent cycles both domestic and foreign consumption must synergistically interact to produce increasing returns. Unfortunately, India's consumption is constrained due to low threshold incomes, poverty and inequality. India's record is poor on these indicators compared to other late industrialising countries (see Table 2.3). Given the decoupling of the Indian industry, we have seen that the linkages to other sectors are not as strong as they could be, and suggest that the ICT/software trajectory as currently displayed does not portend technical change at the economy-wide level.

To understand what sort of learning and innovative capabilities are fostered through this new model of development, it is important to show, theoretically, that outsourcing does not lend to critical services, while technologically and commercially more profitable activities are being farmed out. There are various structural and institutional barriers to global outsourcing of high-value and complex projects. These include: weakness of local competence and a greater desire of MNCs to control technology leakages from foreign subsidiaries, high costs of co-ordination, communication and control of multiple sites; and cultural distance (Mariani, 1999; Maskell and Malmberg, 1999; Narula, 1999; Reddy, 1997). To tap into codified and tacit knowledge, 'face to face interaction' is critical (Cantwell and Santangelo, 1999, p. 104). Socialisation becomes the basis for transferring tacit knowledge, which obviously calls for geographical and cultural proximity (Schmitz, 1999;

Table 2.3 India's economic and social structures in comparative perspective

	Life expectancy in 2000 (years)	**Combined Enrolment Ratio (1999)**	**GDP per capita (PPP US$, 2000)**	**HDI rank (out of 173)**	**% of adult illiteracy (15 and above) 2000)**
S. Korea	74.9	90	17,380	27	14.0
Brazil	67.7	80	7,625	73	14.8
China	70.5	73	3,976	96	15.9
India	63.3	55	2,358	124	42.8

Note: PPP = purchasing power parity; HDI = Human Development Index
Source: UNDP, 2002

Amin and Wilkinson, 1999). Subsidiaries or subcontractors must content themselves with being adjuncts to the central functions of their clients, leading to the decoupling discussed earlier. The local industry is then locked into a low-innovation trajectory, exacerbated by the absence of dynamic domestic demand.

However, as India's export profile shifts with growing software and service exports it is possible to leverage this growth for system-wide economic transformation. Strategically, it calls for an innovation policy that exploits the synergies associated with convergence technologies. The entire gamut of ICT-related convergence technologies becomes a 'carrier' technology, which could transform all other sectors in favourable ways (Dicken, 2003). The world economy has experienced several long booms (Kondratiev cycles) often associated with particular 'carrier' technologies. In the late eighteenth century core economic expansion was driven by textiles, textile chemicals and textile machinery, along with water power. The mid-nineteenth century was driven by steam engines, machine tools, and iron and steel. In the late twentieth century, from the 1980s onward, we witnessed the expansion of computers, electronic machinery, telecommunications, fibre optics, the Internet, software, robotics, information services, digital networks and satellites. Together, these contribute to convergence technologies, which are driving the economic expansion of recent times. However, the synergy among these technologies is generally applicable only to core capitalist economies. If history is any guide, the logic of a previous carrier technology – say, the steam engine, iron and steel, the internal combustion engine, the telephone – has had only a selective impact on most developing countries. Why should convergence technologies have outcomes different from previous carrier technologies? Are there intrinsic differences between the new carrier technology and those of the past?

This is not the place to engage this debate. Suffice to mention that many carrier technologies did not diffuse well in poor economies, including India, because of various impediments, such as high fixed costs. However, today in 2005 the application of convergence technologies is bringing unit costs dramatically down, thereby contributing to the general diffusion process. Furthermore, convergence technologies, including the Internet, are not only sources for information exchange but also platforms for information storage and retrieval in real time. These are qualitatively different features from earlier information-oriented carrier technologies, and they promise an immense range of applications to increase productivity, such as automation in manufacturing and services, agricultural market information, education and health, to name a few. However, that the Indian industry remains locked into its particular business model suggests that far more needs to be done by the government and the industry to make ICT a genuine carrier technology. Not only must all institutions produce and use ICT technologies, but also these technologies must percolate down to the community and household levels.

The economic reforms of the 1990s provided a new institutional environment for the upgrading of the Indian industry. Reforms have opened up the foreign sector as a source for markets, capital and technology. The software industry has taken advantage of new opportunities. India's foreign exchange reserves are at an all-time high. However, rapid growth of export revenues has negative consequences such as the appreciation of the Indian rupee, which could hurt software and other export-driven sectors, such as textiles, that are not as dynamic as the IT industry (see Chapter 5). Since 2004, with comfortable foreign exchange reserves, the Indian rupee has been strengthening against the US dollar (*Business Standard*, 2004). It is clear that India's economic resources are being reallocated today in 2005 based on India's competitive advantage, but it is not clear what the balance sheet looks like in the overall macroeconomic context.

While there are some signs to suggest that poverty may have declined, income distribution has worsened (D'Costa, 2003a; UNDP, 1999). The routing of the incumbent ruling party, the Bharatiya Janata Party, in the election of 2004, which oversaw high growth rates, increased trade, and India's global visibility is indicative of the maldevelopment that has accompanied liberalisation and IT sector development. Such political outcomes suggest the importance of creative social and economic programmes that transfer real resources from the profitable software industry to the broader ICT and non-ICT sectors. These could be in the form of job-generating investments, technological upgrading, research and development, worker re-training and infrastructure development. Without such compensatory measures not only is there a possibility that the Indian industry will run out of steam before meeting its full development potential, but also the government itself could be accused of complicity in creating a dual India.

4. Policies for broad-based development

The Indian IT industry has been shown to be locked into a lower innovative trajectory. Despite geographical clustering of software firms, in

Bangalore and in the Santa Cruz Export Processing Zone in Mumbai, the local agglomeration effects are still weak, as software, unlike manufacturing, does not have the same intensity of backward and forward linkages (Lema and Hesbjerg, 2003). Yet, to be on a higher innovative trajectory, the industry needs to diversify markets and project experience. External links can help, but they also impose constraints on learning a wide range of complex projects. Consequently, firms cannot seize alternative opportunities and get out of the 'same frame of mind' (Hamel and Prahalad, 1994), locking themselves tightly to client-driven, external-driven demand. Learning is easier when socialisation among suppliers and between users and developers is institutionally fostered (von Hippel, 1988). With a weak user base in India, evident from not only low level of ICT goods manufacturing but also low technology diffusion, the decoupling process restricts collective learning. Though complex in design and implementation, it is necessary to build tight linkages between domestic and diversified foreign ICT markets. This would entail generating intense interactions between users in India and abroad and the co-designing of complete software projects in India.

Considerable challenges lie ahead, such as creating and sustaining a huge pool of highly trained technical talent. While world demand for ICTs can be expected to grow, the supply of technical professionals is likely to be strained (D'Costa, 2004b). Some developed countries meet this through limited immigration of foreign IT workers. However, the Indian industry association the National Association of Software and Service Companies (NASSCOM) foresees a large deficit in the supply of technical talent and has been urging the government to intervene. The association recognises that the existing educational infrastructure is 'stretched to the limits' and fears serious problems in the quality of students and teachers (NASSCOM, 2002). Of the 160,000 engineers produced each year in India, only 3000 meet international quality standards. Here is another issue. A knowledge-based economy is not just composed of technical disciplines. It includes a diverse range of talent, from engineers and social scientists to managers to graduates broadly trained in the humanities. Lastly, the NASSCOM figures also show that much of the shortage will arise because of external demand. About 75 per cent of total demand for IT professionals in 2004/2005 is expected to be in the export sector. If this optimistic growth is realised with increased supply of talent, we can assume that decoupling will worsen.

To leverage the IT industry for economic transformation, there are at least three conditions that must be fulfilled. All three are integral to domestic market development. First, sustained expansion of both foreign and domestic markets will be critical. From the demand side there are significant barriers to overcome, such as income inequality and high incidence of poverty. From the supply side, however, firms are best placed to undertake innovative activities. Firms have undertaken varying learning initiatives within the narrow trajectory of software services exports (D'Costa, 2001), but far more could be done. Innovation strategies aimed at learning, cost reduction and higher revenues per employee and masterful applications of software to a highly heterogeneous Indian setting will be the beginning of a carrier technology. Policy-makers must creatively respond to intensifying global competition to remedy the impending shortages of technical talent.

The experience of the Indian software industry is clearly a result of past education policies, favouring tertiary technical education to create a supply of engineering talent for the largely state-led import substitution industrialisation programme of the 1960s and 1970s (D'Costa, 2005). In an economy of scarcity, formal education at the college level has been a vital link to social and economic advancement. The huge Indian bureaucracy and the public sector have relied on a steady supply of such personnel. Those with technical and professional training were rewarded with remunerative, secure employment in public and private enterprises. A self-reinforcing dynamic facilitated the internalisation of the ethos of higher education by Indians, even as most were structurally excluded from the fruits of economic and social development.

Both economic imperatives and the social penchant for higher education contributed to substantial growth in post-secondary educational institutions. For example, there are nearly 10,000 schools (compared to less than 1500 in 1961) above the higher secondary level. Seventy per cent of these are concerned with general education, while the professional and technical schools comprise about 20 per cent. Disproportionately more public resources have been allocated to higher education and professional programmes such as engineering and medicine, fields typically sought after by the middle class. While 79 per cent of the education budget in 1995/96 was allocated to elementary and secondary education, which serves a far larger number of students, 16.4 per cent was allocated to higher and technical education, which serves a small percentage of students (World Bank, 2000). Based on the 1995/96 education budget, this roughly translates to an annual average expenditure of Rs 2875 and Rs 11,307 (roughly US$100 and US$400) for a high school and college graduate respectively. Technical education is likely to be more expensive. What this means is that past education policies have been biased towards the better-off social groups and often at the expense of more rudimentary universal education. Biased policies are no doubt paying off via India's ICT industry, but much remains to be done for the education sector as a whole.

While the narrow ICT development has rested on the abundant absolute supply of technical talent, there has been under-utilisation of talent as well (D'Costa, 2002b). To widen the benefits of ICT and its diffusion, greater spending on mass education is necessary. Fortunately, there is some institutionalised access for the underprivileged groups through the reservations system (affirmative action) for low caste Dalits and 'other backward castes'. But it would be a gross omission if widespread education of the masses and their social welfare were not made integral to the development process. It is attractive to suggest that, despite massive poverty and illiteracy, India has made a dent in a segment of the global high-tech industry. Alternatively, it is equally tempting to suggest that such development was possible at the expense of the poor and the marginalised. For inclusive national development, education policy needs to be reassessed so that the benefits of education do not accrue only to the better-off sections of the population and to particular regions.

India also needs to create talent to meet impending shortages at home. The industry is aware that the global demand for ICT goods and services is both an opportunity and a challenge. It remains to be seen what position India will ultimately occupy in the evolving international division of labour in the production of ICT. In the 1990s, as Indian software firms became more credible with high quality work and costs remained an important arbitrage in their favour, many MNCs either established their own subsidiaries in India or created Indian outsourcing partners. This division of labour, as evident from the decoupling of the Indian software industry from the manufacturing sector, suggests that India (along with a few others) could be positioned in the middle of a triangular relationship: with the US, Western Europe and Japan focused on design activities, India on software development and Japan, East Asia and South-East Asia on hardware, components and peripherals-manufacturing. If we assume that manufacturing in general will be difficult for India, then conceptual activities such as design could be realised with India's availability of technical talent. This will not come about automatically. Rather, the entire institutional environment will need to be re-focused, favouring higher order, design activities based on a solid science and technology foundation.

Since the 1980s a large number of highly successful Indian technical professionals have migrated to the US and other OECD economies.

Tapping non-resident Indians (talent residing abroad) and retaining those that are in India would be a step in the right direction. Transforming international brain drain to national brain gain must be accomplished in addition to reversing the problem of India's internal brain drain (from non-ICT to ICT sectors). Brain drain is related to education and economic opportunities. Since education cannot be deliberately curtailed nor draconian measures imposed to control the movement of people, Indian talent can be retained or brought back from abroad by creating conducive employment and social conditions in India (Saxenian, 2002). The Taiwanese and Chinese cases throw some light on these issues. It is clear from both cases that developing domestic markets and providing a favourable business and employment environment for international competitiveness contribute to national production and foster higher local consumption through increased output and income growth.

India's success in the US ITES-BPO market could also lead to new challenges (D'Costa, 2004b). While perceived and real labour shortages in the US could be met from India, the current technology slump is a reminder that high demand is not always durable. Furthermore, there are many other countries well-endowed with human resources, which could challenge India's first-comer advantage. Russia and the former East European countries, China and the Philippines are potential competitors (*Financial Express*, 2004). Hence, the expansion of the Indian domestic market will be critical not only for its intrinsic developmental impact, but also to help stem the flow of talent in large numbers to other countries. Alternatively, and complementary to the current export pattern, Indian firms will have to diversify their export markets.

One market that needs to be examined is Japan, which currently draws little ICT services from India. Japan is increasingly confronted by demographic challenges such as an ageing population, declining fertility rates and increasing dependency ratio. In order to maintain their lead in the IT-related high technology industries, Japanese firms will be compelled to import foreign technical talent. While there are many countries, especially China, that could meet this need, the Chinese economy itself is becoming heavily dependent on its own talent. Hence, for India there is a great opportunity to re-orient its IT industry. To make the transition successfully both India and Japan will have to collaborate in the broader area of immigration and industry-specific area of technical education.

5. Conclusion

This chapter has examined the trajectory of the Indian ICT industry and its particular form of articulation with the world economy. The industry does not wholly fit any particular model of development. It is not a product of industrial upgrading where older industries and the agricultural sectors yield to new technologies and higher productivity throughout the economy. Rather, it is a form of leapfrogging within the *narrow* ICT sector in the midst of low productivity manufacturing and agriculture sectors.[4] The industry's trajectory is shaped by the cumulative effects of past import substitution industrial and education policies and subsequent global expansion of the ICT industry. India's initial success is now being reproduced on a larger scale, but questions about the depth of the industry's transformative capability of the economy remain, given the subordinate nature of the Indian market to foreign and undiversified markets and the weak linkages to the rest of the economy.

The trajectory associated with the Indian ICT sector does suggest a transition to a new kind of business model, which is consistent with global economic integration fostered by neo-liberal policies. However, the dynamics of the ICT sector illustrate a truncated form of development in which the ICT sector is decoupled in a number of ways: from the broader domestic market; from associated hardware manufacturing; from the more lucrative software package development; and from the Japanese and the larger European markets. This path-dependent, cumulative outcome is a result of institutional history and structural constraints. It is also a policy outcome. The relatively abundant supply of technical skills in an impoverished economy is a result of past policy interventions that were directed at import substitution industrialisation and not exports. Only over time has it become more purposeful and consistent with globalisation. Global demand and international mobility of talent have now reinforced an ICT-specific education supply system. Moreover, the relationship between formal research and development and technical change is nebulous and the Indian industry does not yet display an interest in R&D. This is one example of how received theories of economic development can only be used selectively to explain the rise of the Indian industry. Deregulation as a response to the macroeconomic crisis in India has also boosted the fortunes of the industry mainly because the sector was not strapped with the legacy problems found in other sectors. Firms were small and could flexibly adjust to global demand for simple IT services. Only a handful of Indian firms have moved up the value chain with more complex skill sets.

To make a durable transition from an impoverished economy to one driven by a socially inclusive development process, Indian firms, with the assistance of the state, have to take a leadership role to get more out of new global opportunities rather than be content with foreign subcontracting projects. There are two dimensions to this issue. The first is that India itself must contribute to the world market through its own consumption and production of ICT goods and services. This is a tall order; production requires investment in manufacturing facilities and production infrastructure geared towards a knowledge-based economy, while consumption requires reducing poverty and inequality, raising incomes, increasing government social spending, private spending on research and diffusion of ICT in all modern economic activities. While there are healthy signs regarding increased ICT spending by businesses and the government on hardware, telecommunications, software and IT services (Silicon India, 2004a), more intractable areas such as basic education, training and traditional infrastructure development remain on the sidelines.

The growth of the software industry represents a favourable development in the Indian context, heavily path-dependent and influenced by the external sector. But it is premature to claim that it represents a new 'model' of development, with transformative capability, worthy of uncritical emulation by the wider community of developing countries. The transformative potential of the Indian ICT industry has been overestimated because the serious challenges posed by the four forms of decoupling of software are overlooked. There are, of course, benefits from the current expansion of the industry, but because of its enclave-like characteristics economy-wide technical change, crucial to structural change, has been limited thus far.

The ICT industry presents a (business) model of development, which is strongly linked to the world economy. However, its particular articulation, as evidenced by decoupling, suggests that such links need to be re-evaluated when focusing on the issues of economic development and structural change. The source of the tension arises from the nature of labour in the software industry; shifting costs of labour due to increasing demand, in this case from the external sector, and the price distortions set in motion due to uneven rates of sectoral expansion. Development theories centred on factor prices and their changes due to global integration provide only partial explanations, as they do not account for the local institutional legacies and sector-specific industrial targeting by the state. Not only is there an inherent tension between domestic and foreign sectors when it comes to allocating human capital, but it suggests the limits of the sector in transforming the Indian economy under the current business model.

The business model itself must be changed. While exports will remain the mainstay of the industry for the foreseeable future, oriented to offshore development, significant development of the domestic market should be fostered. Broadly, it could be termed 'recoupling', whereby the development of the hardware industry, product niches and diversification of markets, both domestic and foreign, will be critical to industrial upgrading. Recently Nokia and South Korea-based semiconductor firm Intellect Inc. announced plans to set up cellular phone and semiconductor manufacturing in India, which is expected to contribute critically to the 'recoupling' process (Silicon India, 2004b; Silicon India, 2004c). ICT as a whole can be a carrier technology only if there is widespread diffusion through complex production systems and society-wide ICT consumption.

There are some policies that could re-orient the Indian ICT industry. A more aggressive development policy aimed at mass education and poverty reduction will continue to be important for the diffusion of ICT. Public policies will be necessary to support such outcomes. Equally important is the state's role in contributing to the industry's competitiveness and industrial upgrading. One way this can be accomplished is to make the Indian economy attractive to expatriate Indian talent to invest in start-ups. The government has already set up several technology parks and institutions for technology training. Since highly trained people tend to move to locations offering better economic and professional opportunities, it is imperative that the brain drain problem be turned to India's advantage by gaining the brain it has already lost and retaining talent at home. This will be critical to firm-level innovation strategies and India's repositioning in the international division of labour.

The trajectory of the Indian ICT industry suggests that development stages cannot be skipped in a routine sense but they can be compressed, as the East Asian cases demonstrate. In speeding up the process, the role of the state is significant. Contrary to the proponents of neo-liberal policies, the state has an important role in fostering technological upgrading and increasing labour productivity. These are, after all, the twin drivers of economic development and structural change. Sound educational policies and skill development combined with firm initiatives for learning are consistent with knowledge-based growth. In this transition to a new trajectory, opening up the economy to foreign participation is also important. Some of these elements are already at work, with increasing exports contributing to rising productivity and macroeconomic stability. However, it is amply clear that export-driven development of the software industry has its limits. A single subsector cannot be a carrier technology unless the trajectory is broadened to include related sectors, namely, ICT as a whole and its applications in non-ICT sectors. The challenge is to avoid being blindsided in our haste to acknowledge the spectacular growth of the Indian software industry and underestimate the significance of solving more fundamental national development challenges. Only by recognising this short-sightedness and doing something about it will make ICT propel the transition to a broad-based transformative process.

Notes

1. These figures do not include the inexpensive supercomputers built and exported by the Centre for Development of Advanced Computing (C-DAC) and the Indian Institute of Science (IISc), indicative of the hidden talent that has not been fully utilised by the domestic economy.
2. The high value of exports is attributed to Irish government tax incentives to make MNCs re-export the imported software packages (Ashish Arora, Carnegie Mellon University, 2003, personal communication).
3. ITES are typically call centres and other back-office business process operations (BPOs). These activities rely on information and communications technology (ICT) for handling customer transactions such as credit card billing, medical transcription and insurance claims but do not require highly skilled technical professionals. However, not all ITES are low-end service activities. Medical diagnostics and back-office processes for technology services require highly specialised technical personnel.
4. There are a few exceptions to this prognosis. The auto industry's engagement with incremental innovations during the 1980s and 1990s is finally beginning to pay off (see D'Costa, 2005). However, the industry is nowhere near occupying the global position achieved by the Indian IT industry.

3
The Political Economy of IT-driven Outsourcing

C. P. Chandrasekhar

1. Introduction

Since the late 1990s India has recorded dramatic increases in the aggregate earnings of its software and IT-enabled services (ITES) sectors, so much so that software and ITES are often quoted as the source of a new dynamism characterising the Indian economy. This experience has also been used to argue that the revolution in information technology offers developing countries new opportunities in a dynamic industry. In particular, with export revenues accounting for an overwhelming share of India's aggregate revenues in this area, the Indian experience is also seen as indicative of a new opportunity for services-based, export-led growth in developing countries. The emergence of that opportunity is attributed to two outcomes of the revolution in information and communications technologies (ICTs): first, an acceleration in the process of outsourcing of IT and ITES by corporations in the developed countries; and second, an increase in the extent of offshoring of these outsourced activities to countries like India. As a result of these developments, the ICT revolution is expected to alter the global distribution of incomes at the margin and redress existing international inequalities. Not surprisingly, the outsourcing phenomenon has attracted an excess of attention.

However, notwithstanding the substantial attention devoted to outsourcing, discussions on the subject lack conceptual clarity. In particular, both estimates and analyses of recent trends do not distinguish between outsourcing, offshoring and outlocation. This results in adducing an excessive degree of novelty to current processes and also affects the assessment of the implications of the phenomenon. To start with, the failure to distinguish between outsourcing and offshoring results in an excessive degree of optimism about the opportunity available for developing countries in the new context, and excessive pessimism about the loss of jobs that would accompany the shift of services out of developed countries. Further, the failure to distinguish between offshore outsourcing and outlocation results in an overestimation of the benefits to developing-country firms and economies from the growth of software and ITES export and an underestimation of the benefits that accrue to developed countries and their firms from outlocation.

2. Outsourcing: some conceptual issues

Parrinello offers an adequate and useful definition of outsourcing within the context of how outsourcing leads to unbalanced growth. According to him:

> Outsourcing must be conceived as a substitution of integrated processes, which produce and use an intermediate (material or immaterial) good or a business service, with distinct processes, which supply and use the same intermediate element. If initially this element was not a vendible commodity, its outsourcing brings about both a process innovation and a commodity innovation. In particular, a certain industry may outsource part of its internal service activities that then become the outcome of specialized processes and are recorded as part of the service sector. (Parrinello, 2004, p. 397)

In every context, at each point in time, firms and institutions choose to determine the boundaries of their own operation by selecting the parts of their overall activity (say a fully fledged production, marketing and sales operation) they want to undertake in-house and those which they prefer to leave to other firms or institutions. This would determine the degree of vertical integration in production and the degree of involvement in activities other than production, such as accounting, computerisation, marketing and consumer relationship management.

It should be clear from this discussion that outsourcing *per se* is not a global phenomenon. It can and tends to occur in the first instance within national boundaries, in the form of outsourcing of in-house production of manufactured intermediates, in-house processing of raw materials or in-house execution of a range of services. But once the process of outsourcing a particular activity that is internal to firms in an industry is generalised across firms, the shift of the location of the vendor providing the vendible material or immaterial good is no more a process of outsourcing. Thus if activities outsourced previously now exist as independent activities in the countries of origin of the intermediate demand and are relocated to a developing country such as India, the process is no longer akin to outsourcing but rather comprises offshoring an already existing independent commercial activity. Further, if the relocation involves the current producer of the vendible intermediate establishing a commercial presence in the country to which the activity is being relocated, the process is one of outlocation or relocation.

It is perfectly possible, and it does happen, that the initial outsourcing occurs through the transfer of a separated part of an integrated process to an independent vendor located abroad. This is the only instance where the process broadly categorised as 'outsourcing to India' (say) is an actual example of outsourcing and not just offshoring or outlocation. Thus, seen inter-temporally, what we have is that for each level of cumulative outsourcing, there is some of this activity that is undertaken by firms within the country of origin of the demand and some that is undertaken by firms from outside. It is the implications of the latter that concern us here.

Different kinds of considerations could enter into the outsourcing decision. The first, of course, is the cost consideration. There are many ways in which outsourcing can lead to reduced costs. For example, certain activities are best undertaken by specialised entities, which are able to reduce costs through such specialisation either because of the large scale of operations they can sustain by pooling demands from multiple players or because of their ability to keep overhead costs down and run profitably a small-scale operation. In other cases, outsourcing may reduce costs by transferring activities to firms operating in a labour market segment different from that of the outsourcing firm, where wages and, therefore, the costs of undertaking certain activities are lower.

Needless to say, such factors motivating outsourcing of activities encourage offshoring as well. Thus, the effort at cost reduction can encourage developed-country firms to outsource directly to developing-country vendors, since they have access to a segment of the global reserve army of unemployed labour where the reservation wage is much below that of the developed countries. Further, competition between vendors of outsourced products or services in the developed countries or between vendors in developed and developing countries could encourage the former either to outlocate or outsource a part of their own operations to environments providing access to cheap labour.

Not surprisingly, the greatest advantage of offshoring to India is reportedly the cost savings enjoyed by the offshoring companies, 70 per cent of whom are US-based. Cost comparisons between India and the US point to savings of around 80 per cent accruing to the source companies.[1] Workers in the US can cost firms anywhere from $30 to $35 per hour as compared to around $18 per hour for their Indian counterparts. A study published by the University of California, Berkeley's Fisher Center for Real Estate and Urban Economics in 2003 illustrates how much less a worker in India makes in base wages per hour compared with a worker in the US performing essentially the same job. For example, a worker earning around $35 per hour as a financial analyst in the US, or a worker earning around $23 per hour as an accountant in the US, would receive only from $6 to $15 per hour if working in India. Telephone operators and medical transcribers earning $13 per hour in the US would receive less than $2 per hour in India.[2] According to one Silicon Valley company estimate, the monthly cost of keeping one tech employee in the Valley is $15,000 per worker, whereas someone with the same skills, responsibilities and job package would cost $2500 per month, in New Delhi. This translates into savings of $12,500 per month or $150,000 per year. By sending ten jobs to India, a start-up can thus save $1.5 million of payroll costs.[3] Investment costs are also lower. The cost of setting up a call centre in Delhi is estimated at around a third of the cost in the US.[4]

Besides reduced costs, a second factor that could motivate firms to outsource certain activities can be their desire to convert the fixed managerial salary and wage costs involved in undertaking an activity in-house into recurring costs, allowing them to curtail those costs when capacity utilisation falls due to demand constraints. Finally, firms may outsource because they find the need to focus their managerial attention on core areas, rather than fritter away energies in micro-managing non-core activities.

However, a contrary trend is possible as well. Firms may in particular contexts choose to undertake internally a set of activities that can be outsourced because of factors such as economies of scope, insufficient quality guarantees or supply uncertainties. In sum, managing outsourced activities can involve costs that outweigh the benefits.

Further, there can be external constraints to the ability of a firm to outsource its activities. Mainstream discussions of the boundaries of the firm have noted the existence of activities which cannot be outsourced easily because of 'hold-up' problems.[5] Put simply, if a possible target entity to which some activity of a firm can be outsourced needs to make substantial relation-specific investments to undertake the activity, it is confronted by the fact that the outsourcing contract is an incomplete guarantee that the returns on that investment would be fully realised. The outsourcer may choose to pull out of the contract, well before the investment is amortised and returns obtained. This deters the potential vendor from entering into an outsourcing contract and restrains the extent of outsourcing that occurs.

3. Technological change and outsourcing

The reported increase in outsourcing during the 1990s suggests that changes in circumstances have increased the benefits from outsourcing. This appears to have been true initially of manufacturing and subsequently of services. Technological change is seen as having played an important role in both cases. In the case of manufacturing, the acceleration of offshoring and outlocation, which began in the 1980s, was driven by three features of technological change. First, as has been observed in the literature (Fröbel, Heinrichs and Kreye, 1981), technical change has resulted in the segmentation of production processes, with each segment being characterised by different labour and input requirements. This makes the appropriate location for individual segments, from an input and labour access point of view, different. Second, technological change in the transportation sector has clearly reduced the time required and the costs of transporting commodities over long distances. This allows for the production of parts of a commodity in different global locations and their assembly into the final product in yet another location. Finally, technical changes in the communications industry have reduced costs and facilitated connectivity, enabling and easing centralised management of worldwide production facilities. All of this has favoured global outsourcing.

Thus, Feenstra and Hanson (1996, p. 240) argue that 'the fragmentation of production into discrete activities which are then allocated across countries' has resulted in an internal reorganisation of production within particular industries and a consequent increase in international outsourcing. In particular, confronted with import competition from low-wage countries, firms were responding by shifting non-skill-intensive activities abroad, resulting in a relative increase in the demand for skilled labour in the US. Measuring outsourcing by the share of imported intermediate inputs in total purchases, Feenstra and Hanson (1996) estimated that it had increased from 5.3 per cent of material purchases to 11.6 per cent between 1972 and 1990. Overall, differential wage costs appeared to be an important contributor to US outsourcing, resulting in an increase in non-skill-intensive manufacturing in low-wage countries and a relative increase in skill-intensive activities in high-wage countries.

Feenstra and Hanson (1996, p. 242) found that firms in certain industries showed a much higher propensity to outsource than others. In footwear, electric and electronic machinery, instruments and 'other industries' (jewellery, toys and sports equipment), the share of imported intermediates in material purchases was about 1.5 to 1.75 times the average. This was because of two characteristics of the mostly semi-durable consumer goods that were produced by these industries. First, the production process could be separated into self-contained stages, facilitating the transportation of inputs across space. Second, production stages varied considerably in the relative intensity with which they used labour of different skill-types, which created a rationale for moving non-skill-intensive activities abroad.

What is interesting about Feenstra's and Hanson's results was that often this implied the increase in service-intensive activities in the developed, high-wage economies and assembly operations in the developing countries. Product design and development requires workers with a college education and the production of components may require skilled workers. Product assembly, on the other hand, generally requires only workers with rudimentary skills. The first of these features implies that often a foreign firm may be contracted to manufacture a product which is designed by and will be distributed by a US firm, as is true in the case of athletic shoes 'produced' by Nike.

The net result, however, is that much of the value added is due to activity in the US (Tempest, 1996; Feenstra, 1998). As an example of outsourcing, consider the Barbie doll. The raw material for it comes from Taiwan and Japan, while the doll itself is assembled in Indonesia, Malaysia and China using moulds imported from the US. The cost of Chinese labour (35 cents), raw materials (65 cents) and transportation and overheads, including profits, takes the export value of the doll to $2 when it leaves Hong Kong. However, the doll sells for $10 in the US, of which Mattel earns at least $1 and the rest goes to cover the costs of transportation, marketing, wholesaling and retailing in the US. Thus, much of the value-added accrues in the US (Feenstra, 1998, pp. 35–6).

In sum, manufacturing outsourcing, resulting from the segmentation of production processes and the shift of unskilled labour-intensive activities to developing countries, does not affect the profitability of US firms, but helps shore it up in the face of competition. Nor does it affect value-added and GDP generated in the US. But it does reduce the share of unskilled workers in the total in the US. In the process, it has the danger of pitting poorly paid workers in countries like the US against similar workers in low-wage, developing economies. Finally, since skill-intensive activities like R&D and design are undertaken in the source country, offshoring in manufacturing appears to contribute to a *relative* increase in service sector output and employment in the developed countries.

3.1 The shift to services

The last of these is of relevance because, in the second phase of acceleration of the outsourcing/offshoring phenomenon, there has been a noticeable spread in the direction of services. There could be three reasons for this: first, changes in the organisational structure of firms, which permit the dissociation of a growing set of service activities from the core operations of the entity; second, changes in technology, which permit the remote delivery of an intermediate service, even though its production and supply must occur simultaneously and not sequentially; and third, the growing global homogenisation of skills of certain kinds of service workers, partly because of capitalist development in the peripheral countries characterised by lower wage structures. A combination of these tendencies would result in an increase in the outsourcing and offshoring of services.

To start with, there was a significant degree of in-country outsourcing of services in developed countries like the US before the current offshoring boom began. Thus, a survey conducted in the US by the Outsourcing Institute reportedly found that companies with over $80 million in annual revenues increased outsourcing by 26 per cent in 1997 to $85 billion. IT was the fastest growing activity being outsourced, accounting for 30 per cent of total outsourcing expenditures. Human resources was the second largest (16 per cent), followed by marketing/sales (14 per cent) and finance (11 per cent). Manufacturers accounted for nearly two-thirds of the outsourcing, with information and professional services each accounting for 13 per cent of the total (OECD, 2000c).

The offshoring controversy relates to the shift of a part of outsourced services activities to developing countries like India or to the direct outsourcing abroad of a new set of activities. This too has been enabled by technology, especially ICTs. Advances in ICTs and the consequent fall in the costs of data generation, storage, transmission and dissemination have allowed for the remote delivery of a range of services. This change has resulted in an increase in cost-border outsourcing of services. Prior to the digital revolution's transformation of service activity, the provision of most services required the presence of a service provider at the point of delivery of the service. As a result, services export took the form of migration of personnel to the location where the service was provided, as epitomised by the migration of skilled technicians, doctors and nurses to the US and of semi-skilled and unskilled workers, including carpenters, masons, chauffeurs and housemaids to the Arabian Gulf countries from India. Benefits to the home country came in the form of *remittances* of hard currency earnings by these migrants to their families, which augmented the scarce pool of foreign exchange available to these countries. But the magnitude of such income was limited by the restrictions on the movement of skilled and semi-skilled and unskilled personnel set by the immigration laws and practices of countries where the relevant service demand originated. In the IT services area, this form of delivery was reflected in the use of H-1B visas to provide IT services onsite in the US, India's principal market.

The digital revolution has helped to overcome this obstacle. Now there is a range of services being provided by workers located in a country different from the one in which the service is actually delivered. These services are *outlocated* or *outsourced* and delivered via telecommunication or the Internet from remote locations like India's metropolises. These changes, while driven by substantial savings in wage costs, also imply that managerial costs of offshoring have been reduced, and that the difficulty of finding suitable vendors has fallen substantially. The evolution of technology and the increase in the spatial spread of potential vendors have made it easier to find vendors for activities that firms seek to outsource, despite the hold-up problem noted above. As a result, firms tend over time to move activities beyond their boundaries to producers outside, at locations where transportation costs do not neutralise any net pecuniary benefit the firms may derive from outsourcing. Once the process begins, competition drives all firms in an industry to adopt a similar strategy – cost-reducing organisational changes have this in common with technical change.

It should be clear that the outsourcing of segments of a manufacturing process and the outsourcing of service activities undertaken by manufacturing firms are two alternative ways of decomposition of an initial integrated process, in which a good in progress is produced by means of two processes, connected by the usual serial input–output relation. According to Parrinello, 'In both cases the choice between a certain type of outsourcing and other methods of production, should be assumed to be a cost-minimizing choice of techniques and of organization and to imply a certain geographical dispersion of specialized activities' (2004, p. 397). It is no doubt true that improvements in transportation and ICTs render both kinds of outsourcing more profitable than the integrated process, by decreasing the costs of transportation and the costs of communication and data processing. However, while improvements in transportation technology may bias the process of outsourcing in favour of manufacturing (involving material and heavy goods), those in ICTs would encourage more outsourcing of services and immaterial goods, given the need for simultaneous interaction between the provider and the user of services, and the informative content of the immaterial goods.

The ICT revolution, it is argued, helps transform a service produced for a single or few consumers to one that is produced for mass consumption. This allows the service 'industry' to exploit economies of scale just as manufacturing has been doing ever since the Industrial Revolution. A service, such as an online database, produced and placed on the Internet, can be accessed by a large number of consumers. The ICT revolution also changes the relationship between providers and consumers, with the latter being able to access services like health, banking and financial services and entertainment without personal onsite contact with the provider. Easier access, which implies easier delivery, also allows for growing differentiation and rapid diversification of the services 'products' offered to consumers. These changes suggest that ICT both creates new 'services-intensive' products, the production of which can be outsourced and outlocated to developing countries, and helps 'produce' some traditional services at one location (a developing country), remotely providing them at another (a developed country).

It must be noted that what is referred to when we speak of outsourcing and offshoring of IT and ITES is not only services *per se*. We are partly speaking of the process of production of intangible goods such as software, which have been segmented and some segments outsourced. As has been noted, often services increasingly resemble commodities. The conventional idea that services differ from commodities because they cannot be stored, that they have to be consumed at the point of production and that their consumption requires the direct interaction of service providers and consumers is no more true of a range of services. To quote an OECD study:

> Copies of movies and most other performances can be recorded and mass-produced for future consumption, like manufactured products. Software is developed and boxed like any other manufactured

> product, and is considered, for all intents and purposes, a good – albeit with a high service-related content. In these instances services have, in a sense, taken on the characteristics of commodities – one provider is mass-producing a common product for many people. (OECD, 2000c, p. 7)

Thus, the outsourcing and offshoring of services is all the greater because the services sector has substantially changed its character.

3.2 Offshoring and outlocation

What is of relevance to our discussion, however, is not just that changes in technology had resulted in an increase in the extent of outsourcing. More crucially they have, as suggested by the above discussion, resulted in an increase in the extent of offshoring and outlocation. There are two features of the offshoring/outlocation process that need to be noted here. To start with, these processes are not akin to foreign firms' establishing capacities in host developing countries to produce for local markets. What is involved is the transfer to the host country of the production of a good or service which was originally being undertaken in the outsourcing country or in a third country, with the express intent of export to the outsourcing country or a third country. Further, while both offshoring and outlocation imply new employment in the host country, the value-added benefits to the host are greater in the case of offshoring because the profits from local operations (even if based on lower costs and prices) are not appropriated by foreign firms and subsequently repatriated.

4. The global outsourcing market

How large is the global outsourcing market in the IT and IT-enabled services? The difficulty here is that outsourcing as a business practice is itself relatively recent and outsourcing or offshoring to developing countries so novel that assessments of the size of the market are all in the nature of predictions or projections. These predictions and projections themselves are made by international consulting firms, like Forrester, Gartner and IDC, for sale at exorbitant prices to companies looking for ways to cut costs and restructure their operations or to those looking for new markets to venture into. In either case, unless the reports are identifying new opportunities they are unlikely to find buyers. Not surprisingly, selected snippets of information, providing hints of a huge opportunity, are offered to the press to serve as teasers that would persuade them to acquire the full study or attend an expensive, executive education workshop to discuss its results.

As a result, there are many commonly quoted estimates of the IT services outsourcing and the ITES-BPO (business process outsourcing) markets and projections of their growth. These often do not make clear which segment of the market is being spoken of. Partly, as a result of this, they also differ substantially. Gartner is quoted as estimating in 2002 that the global BPO market would grow to $234 billion by 2004. In June 2003, Gartner reported that the worldwide BPO market was expected to grow by 10.5 percent, to $122 billion in that year, up from $110 billion in 2002. Add on ITES, and a Gartner estimate quoted by NASSCOM sees the global market rising from $570 billion in 2002 to $1200 billion in 2006.

But estimates are not uniformly optimistic. Leading technology-impact forecasting firm Forrester Research, on the other hand, reportedly predicted that the core BPO market will grow to only $145 billion, even by 2008.[6] In an August 2003 report titled 'BPO's Fragmented Future', Forrester declared talk of a BPO market bonanza a myth. This was because firms looking to outsource core business processes like human resources and finance and administration because of significant cost savings were reporting not only inability to find a single vendor who could manage such complex offerings, but also being stalled by inflexible contracts, difficulty managing vendors and a lack of performance metrics.

As a result, the market in Forrester's view was likely to fragment into four segments: simple bulk transactions, such as credit card or stock trade processing (estimated at $58 billion in 2008); broad shared services, such as finance and administration, indirect procurement and HR ($57 billion); high-volume vertical processes, including policy administration, claims and loan process applications ($6 billion); and niche vertical applications, like environmental data reporting and chemical process control monitoring ($24 billion). Of these, the first is the easiest to master and is likely to be, along with the second, the area into which developing countries like India are expected to enter. But these would be the areas in which competition would result in a substantial squeeze in revenues per employee, leading to a reduction in wages and/or a squeeze of profit margins. These are global estimates of contracts that are outsourced both at home and across borders. What matters for India and other similarly placed countries is how much of this business will be on offer in the cross-border market.

5. Outsourcing: the Indian example

Estimates made by the National Association of Software and Service Companies (NASSCOM) suggest that revenues from software and IT services increased from Rs 108.99 billion in 1997/98 to Rs 378.40 billion in 2000/2001 and Rs 708.60 billion in 2003/2004.[7] As a result, the share of software and services in the total IT market rose from 59 to 67 and 79 per cent respectively in those years.

The growing importance of software and IT services in the total IT revenues was the result of a rapid expansion in the export of software and services from Rs 65.30 billion in 1997/98 to Rs 283.50 billion in 2000/2001 and Rs 555.10 billion in 2003/2004. The share of exports in the total revenues of this segment of the industry rose from 60 to 75 and 78 per cent respectively in these years.

Exports of IT-enabled services increased from Rs 24 billion in 1999/2000 to Rs 42.50 billion in 2000/2001, Rs 71 billion in 2001/2002, Rs 113 billion in 2002/2003 and Rs 163 billion in 2003/2004. This increased the share of exports of ITES in total software and services exports from 14.3 per cent in 2000/2001 to 19.6 per cent in 2001/2002 and 29.5 per cent in 2003/2004. If this trend persists, ITES would come to dominate IT industry revenues and exports (see Table 3.1).

Global and domestic trends are expected to result in an intensification of these tendencies. For example, according to one estimate, offshore BPO represents only 1.5 per cent of the total BPO market.[8] This could be seen as pointing to the space still available to Indian companies for growth. However, the figures are disappointing when we note

Table 3.1 The structure of IT production and exports in India

Year	Ratio of IT software and services to total IT market (%)	Ratio of IT software and services exports to total revenues (%)	Ratio of ITES exports to software and services exports (%)
1997–98	58.47	59.91	N/A
1998–99	66.70	64.81	N/A
1999–00	66.28	71.52	14.26
2000–01	66.86	74.92	14.96
2001–02	75.25	76.79	19.55
2002–03	77.76	77.52	24.10
2003–04E	79.39	78.34	29.51

Source: www.nasscom.org, accessed 30 January 2005

that even India's current presence in global outsourcing implies its domination of the market, at 80 per cent of the total according to an estimate from *Le Monde* quoted by UNCTAD (2003). If that market is unlikely to grow rapidly, then the space available for new players is indeed limited, unless the market is redistributed. That could mean that BPO is no great opportunity even for a country like India. It would also mean that, as in the case of primary products, competition between developing country providers of lower end services would drive prices down and transfer the benefits of low wages to international corporations, leaving little behind in the developing world.

5.1 Explaining India's success

The point is that even in the model country, namely India, BPO-ITES is only an opportunity in the making. This comes through from figures quoted by NASSCOM, the industry body that is gung-ho about the potential in this area and has (unfortunately) emerged as the leading purveyor of information in this regard. NASSCOM estimates that in 2003/2004 IT software and services yielded revenues of Rs 708.60 billion or about 2.6 per cent of India's GDP. Direct employment created by the ITES-BPO sector is expected to touch just 245,500 people in 2003/2004, up from 42,000 in 1999/2000 (see Table 3.2).

The reasons for offshoring to India are obvious. It is an excellent location in terms of the availability of manpower with the requisite skills (say, basic computer literacy) and the necessary characteristics (for example, knowledge of English). It is also a low cost location with access to cheap, skilled and highly qualified labour. India's competitive advantage in the software export area is also substantial. Even before the downturn of the late 1990s, wage costs in India had been estimated at

Table 3.2 Manpower by horizontals (2003/2004)

Service Lines	Number of people
Customer Care	95,000
Finance	40,000
Human resources	3,500
Receivables management	21,000
Administration	40,000
Content management	46,000
Total	245,500

Source: www.nasscom.org, accessed 30 January 2005

one-third to one-fifth of US levels for comparable work (see Tables 3.3 and 3.4).

As a result of the visible advantages of countries like India and the observed rapid increase in IT services and ITES outsourcing, there is much optimism about the benefits that the new trend can bring to India and other similarly placed countries. The strategy driven by that optimism implicitly sees growth in services as more likely to deliver employment, income and export revenue increases than that in the commodity producing sectors. As has been repeatedly observed, leading Indian IT companies like TCS, Infosys and Wipro are rapidly overtaking the giants of Indian manufacturing such as Telco, Hindalco and L&T when it comes to the number of people employed. The manufacturing sector in India seems to be witnessing its own version of jobless growth as companies shed workers to gain cost efficiencies and automation of menial

Table 3.3 Salaries[9] of software professionals in the United States and India

Category	USA (US$ per annum)	India (US$ per annum)[10]
Help-desk support technician	25,000–35,000	5,400–7,000
Programmer	32,500–39,000	2,200–2,900
Network administrator	36,000–55,000	15,700–19,200
Programmer analyst	39,000–50,000	5,400–7,000
Systems analyst	46,000–57,500	8,200–10,700
Software developer	49,000–67,500	15,700–19,200
Database administrator	54,000–67,500	15,700–19,200

Source: INFAC, Bombay quoted in OECD, 2000a

Table 3.4 Comparison of annual wages in software industry

Country	Computer Programmer		Systems Analyst	
India	4,002	100	5,444	100
USA	46,600	1164	61,200	1124
Japan	51,731	1293	64,519	1185
Germany	54,075	1351	65,107	1196
France	45,431	1135	71,163	1307
Britain	31,247	781	51,488	1287
Hong Kong	34,615	865	63,462	1166
Mexico	26,078	652	35,851	658

Source: Joseph, 2002

jobs drives down the need for more workers. On the other hand, IT and ITES companies are hiring and leveraging technology to grab the worldwide outsourcing market.

6. How sustainable is the Indian outsourcing boom?

What is ignored by this argument, however, are questions relating to the benefit to the host country from the recent boom and the sustainability of the boom itself. There are, in fact, a number of grounds for caution. To start with, India's operations in the ITES area are characterised by the increasing dominance of one area, Customer Care, which accounted for close to one-third of BPO revenues in 2003/2004 (see Table 3.5).

Further, as Table 3.6 shows, while the ITES sector is diversified, excepting medical transcription, which records abysmally low revenue per worker, there is not too much difference in the revenue per worker, which averaged Rs 600,000 (or around $12,200) per year. This points to the fact that low wages do drive the industry. As countries like the Philippines and even China seek to enter this market, the possibility of a profit squeeze in the ITES area cannot be ruled out. What seems to be likely is that, as in the case of manufactured exports, a few developing countries would account for an overwhelming share of total exports and even limited spread could be accompanied by a race to the bottom.

This implies that the real beneficiaries from the offshoring movement are the offshoring firms themselves, which reap substantial cost-reduction benefits and profits. According to a report from the San Francisco-based McKinsey Global Institute (2003, p. 9), for every dollar spent on offshoring by the US in 2002, the total value derived by the

Table 3.5 BPO revenues by horizontals

Service lines	2002–03 ($million)	2003–04 ($million)
Customer care	810	1200
Finance	510	820
HR	45	70
Receivables management	210	430
Administration	310	540
Content development	465	520
Total	2350	3580

Source: http://www.nasscom.org, accessed 30 January 2005

Table 3.6 Employment and revenues from IT-enabled services

	1999–2000		2000–01		2001–02	
	Employed	**Revenue (Rs million)**	**Employed**	**Revenue (Rs million)**	**Employed**	**Revenue (Rs million)**
Customer interaction services (include call centres)	8,600	4,000	16,000	8,500	33,800	17,900
Back office operations/revenue accounting/data entry/data conversion including finance & accountancy/HR services	15,000	9,500	19,000	13,500	35,000	28,500
Transcription/translation services	5,000	1,200	6,000	1,600	6,200	1,500
Content development/animation/ engineering & design/GIS	15,000	8,200	27,000	16,000	30,000	21,000
Other services including remote education, data search, market research, network, consultancy & management	1,400	1,100	2,000	1,400	3,000	2,100
Total	45,000	24,000	70,000	41,000	108,000	71,000

Source: http://www.nasscom.org, accessed 30 January 2005

economy was \$1.4 to \$1.47. While 78 per cent of the value, \$1.12 to \$1.14, was retained in the US, only 22 per cent, \$0.33, was accrued to offshoring destinations like India. Giving a breakdown of the value created in the US by offshoring, the report said that every dollar resulted in savings of 58 cents to US investors and customers. At the same time, it resulted in an increased import of US goods and services by providers in India.

The same report (p. 7) also argued: 'For every dollar of spend offshored, offshore service providers buy an additional five cents worth of goods and services from the US economy, thereby creating exports and extra revenue for the US economy. Providers in low-wage countries require US computers, telecommunications equipment and other hardware and software equipment. In addition, they also procure legal, financial and marketing services from the US.' Several providers in the US offshoring market are incorporated in the US. The report continues, 'These companies repatriate their earnings back to the US, which amounts to an additional 4 cents out of every dollar of spend offshored' (p. 9).

Finally, the report (p. 9) argues that labour freed due to offshoring adds another 45 to 47 cents of value. The total value retained in the US thus adds up to \$1.12 to \$1.14. The report continues, 'Offshoring creates net additional value for the US economy that did not exist before, a full 12 to 14 cents on every dollar offshored' (p. 9). In contrast, offshoring destinations like India capture only 33 cents out of every dollar offshored. Of this, one cent goes to the state government, including sales tax on the supplier industries and supply of power. Another three cents go to the central government in the form of income tax and corporate tax. While the labour employed gets ten cents, profits retained in India add up to another ten cents. The suppliers only get nine cents. Thus unless India hopes to or manages to garner an ever larger share of each dollar offshored the projected growths are far-fetched.

Not surprisingly, according to the Cato Institute, globalisation of services has created a trade surplus in IT services for the US. US exports rose from \$2.4 billion in 1995 to \$5.4 billion dollars in 2002 in computers and data-processing services, while imports rose from \$0.3 billion to \$1.2 billion. Thus the US trade surplus in these services has expanded from \$2.1 billion to \$4.2 billion (Lindsey, 2004, p. 8).

What is more, over time there is growing evidence of offshore activities being set up or taken over by firms from the developed countries.[11] For example, in India captive offshore back-office operations of foreign enterprises like GE, Amex, HSBC, AOL and so on play an important role, though many captives like American Express, Citibank, HP, Dell and AOL also outsource substantially to third parties. Many captives like British Airways have also turned into third parties, such as WNS.

According to estimates by Dataquest, India, the Rs 178.30 billion Indian BPO pie in 2003/2004 primarily comprised four segments: the multinational corporation (MNC) captives, who contributed around 56 per cent of the total; the India-centric third party BPO service providers, with around 29 per cent; the MNC BPOs, with around 8 per cent; and domestic call centres that were primarily captives, which accounted for the remaining 7 per cent.[12]

Acquisition has also become quite common. Recently, IBM acquired Daksh eServices, one of the large and successful Indian BPO service providers.[13] This is a step further in IBM's ongoing efforts to strengthen its India delivery capabilities. Daksh will certainly increase IBM's capabilities in the offshore contact centre market, since 80 per cent of its revenue comes from voice and web support. Daksh focuses on e-commerce – its first client was amazon.com – and the retail, consumer electronics and telecom verticals. According to Gartner, the Daksh acquisition raises serious questions about the viability of Indian-owned business process outsourcing providers. In its view, managing rapid growth is proving to be a big challenge for many 'local' Indian entrepreneurs. As a result, many MNCs are making major moves into offshore outsourcing, and this raises questions about how many 'pure-play' Indian providers can survive their pressure. Established IT service providers like Infosys, Wipro, TCS and HCL Technologies have been able to develop their skills in these areas over many years. For the new players, human resources issues like attrition and security are reportedly emerging as serious management challenges. According to Gartner, enterprises considering pure-play offshore BPO providers must recognise that offshore outsourcing requires a long-term commitment from both parties involved. As the offshore BPO market accelerates and consolidates, size and management experience will become critical decision factors.

Further, the offshoring trend seems to be partially reversing itself. On 26 March 2004, Bharti, an Indian telecommunications provider, announced a major hardware, software and IT services outsourcing agreement with IBM. Under the terms of the 10-year, $700–750 million agreement, IBM will:

- manage all of Bharti's customer-facing IT applications, including billing, customer relationship management and data warehousing

- support Bharti's internal applications, including Internet services, e-mail and online collaboration
- manage Bharti's IT infrastructure
- provide disaster recovery services for Bharti
- support Bharti in delivering voice, data and content-based services

Similarly, Dabur India Limited has signed the US consulting major Accenture to outsource its IT infrastructure and application management functions. Under the ten-year agreement,

- Accenture would manage Dabur's IT functions, including applications management, and provide consulting assistance on the company's business plans
- Accenture would also design, build and run IT systems for the supply chain and secondary sales functions. The IT employees of Dabur were shifted to Accenture with consistent terms of employment.

These tendencies together imply that a large share of the value-added benefits from offshoring go to firms from the developed countries and, as a result, the implications for developed-country GDP would, as in the case of manufacturing outsourcing, prove positive.

There are also signs of weakness in the IT services area. There are three broad categories into which such services fall: hardware services, including maintenance of the growing installed-base of computers; educational services, which generate the skilled and semi-skilled personnel needed to sustain IT sector growth; and software services, including the production of software packages and generation of customised software. In the domestic market, revenue from software services, at Rs 52.69 billion in 2003/2004, was only about one-and-a-half times the combined revenue from hardware services totalling Rs 31.42 billion. That is, not only were domestic service revenues small, but revenues from areas such as facilities management and maintenance generated as much as 60 per cent of the revenue garnered from the production of IT-enabling software. This is of significance because a large domestic software services market provides the workplaces for training for software personnel and the base for generating the skills necessary to meet the higher demands of the export market and to graduate to the higher end of the software value chain. That base is crucial since a substantial part of revenues from IT-related educational services is known to accrue to teaching shops that do not develop adequate skills among their students. A growing disproportionality between the markets for domestic and export software services, therefore, constrains the ability of the system to generate the wherewithal in terms of the personnel needed to service the export market. This is the first factor to take into account when making judgements of India's potential in the export of software services.

Further, to return to a point made earlier, the ability of the IT sector to 'unleash growth that will change India's economic fortunes' depends largely on revenues from exports of services. However, there are three prerequisites for the potential suggested by the factors discussed earlier to realise themselves. First, Indian software exporters need to diversify in terms of sources and destinations. Of the 1250 companies exporting software services in 1999/2000, those exporting more than Rs 1000 million (about $22.5 million) stood at just 37. The top 25 exporters accounted for 61 per cent of export revenues. And the US market dominated in terms of destination, accounting for 62 per cent of exports as compared with Europe's 23.5 per cent. Second, Indian software service providers should be able to sustain the quality of services offered by inducting appropriately qualified and skilled personnel to not merely write code but also design systems. Third, Indian firms should be able to migrate up the value chain, so as to ensure a growing share of the market as well as enter into segments that offer higher value per employee.

In all these areas the availability of personnel is bound to prove a constraint. The inadequacy of training services resulting from the proliferation of poorly staffed, profit-hungry teaching shops and the limited base for training and skill development in the domestic software services segment have already been noted. As a result, even as an outsourcer India still remains a lower-end software supplier and supplier of IT-enabled services. Unlike in the domestic market, both generation of customised software and generating code for systems specified by clients dominate the export software services market. As has been repeatedly emphasised, this often involves some body-shopping in the form of temporary export of software professionals to undertake specific jobs in large projects designed and executed in the West and enclave-type offshore supply.

What is noteworthy is that even the big exporters obtained little by way of revenues from frontline software products or higher-end consultancy and software generation services. According to industry insiders like Narayana Murthy of Infosys, the move up the value chain has hardly occurred and is not India's priority. In an interview Murthy said Indian software expertise in customised services had a long way to go in quantity and quality before focusing entirely on other fields. To quote Murthy, 'Yes, moving up the value chain is a good idea. We are at it ourselves – about Rs 20 crores [Rs 200 million], which is just 8 per cent of our total business' (quoted in Mohan, 1998). There are others, like Vinay Deshpande of NCore (quoted in Mohan, 1998), who feel that while software services should not be sneered at, the contract should be properly designed. Deshpande feels technology thus acquired could then be leveraged to develop indigenous products, for, in the long run, the money is in developing products.

Further, it is clear that a growing share of this export revenue is little more than the sale of cheap skilled and not-so-skilled IT-enabled labour services whose output is transmitted via modern communication technologies to sites where those services are required. A large part of software exports is not very different from the exports of nursing, carpentry, masonry and other such services, except for the fact that, unlike those exports, the presence of the service provider at the point of sale is not required in the case of ITES. The possibility of such service delivery has helped India circumvent the obstacle to service exports created by immigration laws in the developed countries.

6.1 Barriers to entry into software

The difficulty here is that the move up the value chain may not be a matter of pure choice, but structurally limited. While there have been instances of Indian companies delivering high-end products, like the banking and e-commerce software product BankAway, from Infosys, the industry generally accepts that much of the exports from India consist of low-end outsourcing and IT-enabled services. This limited success in terms of the composition of exports may be because there are indeed barriers to entry into higher-end software, resulting from the fact of increasing returns in the form of sharply reducing costs as volumes increase, helping to create and strengthen oligopolistic positions.

What could be the source of barriers to entry into the software sector? Consider, for example, the packaged software segment, mass-producing branded products for large national and international markets. Knowledge products like software packages have the public good characteristic of non-rivalry in consumption, since making the product available to one user does not preclude its availability without much additional cost to another. This essentially implies that so long as the producer and seller of that knowledge can ensure excludability, or that sale of the product to one does not lead to the replication of that package for free use by another, there are substantial increasing returns in the software sector. Producing the first unit of a software product requires large investments in its generation, whereas producing an additional unit is almost costless. The larger the sales, therefore, the lower the average cost and the higher the return.

But that is not all. When large sales imply a large share of the market as well, scale becomes a means of ensuring consumer loyalty and strengthening oligopolistic positions. This is the result of 'network externalities' stemming from three sources. First, consumers become accustomed to the user interface of the product concerned and are loath to shift to an alternative product which involves some 'learning' before the features of the product can be exploited in full. Second, the larger the number of users of a particular product, the greater the compatibility of each user's files with the software available to others, and the greater the degree to which files can be shared. The importance of this in an increasingly networked environment is obvious. Finally, all successful products have a large number of third-party software generators developing supporting software tools or 'plug-ins', since the applications program interface of the original software in question also becomes a kind of industry standard, increasing the versatility of the product in question without much additional cost to the supplier. These 'network externalities' help suppliers of a successful software package to 'lock-in' consumers as well as third-party developers and vendors, leading to substantial barriers to entry.

Thus, conceptually, India's software thrust of the 1990s is not as spectacular as it appears. It comprises substantially export of lower-end software and ITES facilitated by the availability of cheap skilled labour. And it is in large part a technology-aided extension of the earlier waves of migration by service-providers of different descriptions: doctors, nurses and blue-collared workers of various kinds. An expansion of that kind cannot be self-sustaining. Even in quantitative terms the latter development is not spectacular. The 'net foreign exchange revenue' to the country from migration of the old kind, captured by the volume of remittances into India, was in the range of $18 billion in 2003/2004. The gross foreign exchange revenue from software exports is, even after the boom, placed at $12 billion.

7. Conclusion and discussion

Besides the uncertainty created by all these factors, the backlash in the developed countries, especially the US, is threatening the Indian boom. It should be clear that the process of offshoring has adverse first-order employment implications in the developed countries and positive first-order implications in the developing countries. But just as the extent of the benefit to host countries should not be exaggerated, so the loss to the developed countries should not be hyped. According to the NASSCOM annual industry survey, the IT software and services industry was estimated to employ 650,000 IT professionals in March 2003. Of the total, almost 205,000 were working in the IT software exports industry; 160,000 were employed in ITES; 25,000 in the domestic software market; and over 260,000 in user organisations. This is a small share of even organised sector employment in India, placed at 27.8 million in 2001, and almost negligible relative to the aggregate workforce.

How significant has the employment loss in the US been? In November 2003, the premier information technology body in the US, Information Technology Association of America, released an assessment that not more than 7 to 9 per cent of all IT jobs would move out of the US in the next 10 to 15 years.[14]

Contrary to media reports on the rising trend of American companies' resorting to outsourcing and moving jobs offshore, a 2003 report by Forrester Research found that few were actually doing so. The study showed that 60 per cent of Fortune 1000 companies were doing nothing at all with offshore outsourcing, and that 25 to 30 per cent of companies spent only 1 to 5 per cent of their total IT services budgets overseas. Forrester identified these companies as 'bystanders'. The 'experimenters' comprise another 25 to 30 per cent of the Fortune 1000. This segment had experience and relationships with offshore vendors, but offshore was not a key element of their overall strategy, as just one to five per cent of their IT services budget went towards offshore ventures. The 'committeds' and 'full exploiters' made up the remaining 10 to 20 per cent of the Fortune 1000. These firms committed anywhere between 10 and 50 per cent of their services budgets to offshore ventures. Thus the level of participation in offshoring was relatively limited, despite periodic reports of companies' cutting jobs in the US and transferring them abroad.

However, given the declining responsiveness of employment to output growth in the US, the reaction to even this limited degree of offshoring has been aggressive. Unions have released estimates of likely job losses that are huge. And politicians, faced with such a response, have been quick to take up the issue. Despite the loss of many IT jobs in the US following the 2000 crash of the technology sector, Congress had increased the annual allocation of H-1B visas from 65,000 in 1997 to 195,000 in 2003. This was again brought down to 65,000 in 2004. Further, legislators initiated a flurry of bills to deal with the problem. Besides New Jersey, several states, such as Colorado, Wisconsin, Indiana and Minnesota, introduced legislation to ban the offshoring of government contracts. Most of these bills introduced in the state legislatures had not garnered enough votes for approval. Despite the public outcry over the loss of jobs in IT, this lack of approval implicitly demonstrates the support US corporations enjoy among the legislators. However, the US government finally buckled under public pressure and that created by their Democratic opponents before the last elections, and in January 2004 passed a law banning the outsourcing of federal contracts by US companies to third-world destinations.

Though the Bush administration and US corporations are known to be in favour of outsourcing, this backlash against the phenomenon is unlikely to go away. And there is a strong possibility that the clamour will die down only when the issue is resolved through a slowing of the process of offshoring. The reasons for this need to be spelt out. We must begin with a recognition of the fact that a feature of contemporary capitalism is the dominance of finance capital in the US and the other developed countries and the sway of finance across the global economy. The opportunities for profit have resulted in relatively high rates of profit for finance capital, and expectations of high, even if not similar rates, on the part of those investing in the production of goods and services. Ensuring such profits requires a disciplined workforce and limits on wage and salary increases.

In the final analysis, outsourcing and offshoring to reduce wage costs is a way in which corporations exploit and even reproduce the 'reserve army of labour' that capitalism needs both to discipline the work force and keep wages from increasing faster than productivity. If the latter does happen, either inflation will occur or profits will be squeezed. In the current era in 2005, when finance capital dominates in the US and the other developed countries, inflation is anathema, since it erodes the real value of financial assets. On the other hand, capitalists investing in the production of goods and services would be unwilling to accept a profit squeeze. They too need to keep profits buoyant.

The original way in which capital attempted to deal with these problems is through demanding and implementing a certain degree of flexibilisation of the workforce, which helped to impose a degree of discipline through the constant presence of the threat of unemployment. But given the downward rigidity of historically given wage and salary levels and expectations of periodic wage increases, this did not help resolve the profit problem faced by US corporations.

Outsourcing was in the first instance an effort to exploit segments of the labour market where wages were lower because they tapped the reserve army to shore up profitability. When that proves inadequate and opportunities exist or arise to exploit the global reserve army, including that in developing countries like India, offshoring gains momentum. But this cannot be a never-ending process. Either union pressure in the developed countries must stop the process, or the unions have to succumb to it and accept lower wages and meet, even if partly, the profit aspirations of contemporary capital. Thus there must be limits to the offshoring process, not in the sense that it must end, but that it must at least slow down.

Thus the backlash in the US, the as yet limited size of the offshoring phenomenon and the possible slowing of the process of offshoring raise questions about the premises underlying the 'BPO as development opportunity' argument. One is that the Indian experience up to the time of writing can be replicated on the same scale in other developing countries, if supported with governmental and international assistance. The other is that the Indian experience marks a significant development advance and is sustainable. However, the hype surrounding the Indian experience is based more on growth rates from a small base than on the aggregate figures and their relation to economy-wide estimates. It has arisen also because the tendencies triggering the offshoring 'boom' have immanent in them the basis for their own termination.

On the other hand, in the developing countries, the optimism generated by the boom in IT services allows the government to ignore the fact that growth of employment in the commodity-producing sectors has not merely decelerated sharply but is increasingly less responsive to increases in output – the jobless growth syndrome. Needless to say, growth in IT services employment is relevant only to those capable of finding employment in the organised sector, and even in that sector the share of IT services continues to represent a small proportion. The optimism that IT services generate exists only because this is the only segment where employment is increasing significantly. But that growth may be inadequate for most of the population, except the middle-class minority.

Notes

1. See the report at http://www.rediff.com/money/2004/feb/05bpo1.htm, accessed June 2004.
2. US outsourcing threatens one in six jobs in Silicon Valley, at http://www.wsws.org/articles/2004/may2004/sili-m05_prn.shtml, accessed June 2004.

3. See http://timesofindia.indiatimes.com/articleshow/609538.cms, accessed June 2004.
4. See http://www.nasscom.org/artdisplay.asp?Art_id=461, accessed 30 January 2005.
5. For an analytical discussion of this problem, see Holmstrom and Roberts, 1998.
6. *The Hindu*, Saturday, 6 December 2003.
7. Data from www.nasscom.org, accessed 29 January 2005.
8. See http://www.silicon.com/management/itdirector/0,39024673, 10005058,00.htm, accessed June 2004.
9. Figures are starting salaries for large establishments employing more than 50 software professionals. They may be marginally lower for smaller firms. Salaries for a particular designation vary owing to factors such as educational and experience profile of the professional; platform of operation; nature of the assignment (contract or full-time); location of the employer; and additional technical or professional certification.
10. Converted at exchange rate of Rs 41.50 for one USD.
11. When categorised by origin and focus, it is possible to identify four types of ITES: in-house or captive centres; units that were originally spin-offs; business process outsourcing (BPO) units; and broad-based service providers, who offer consulting or IT services in addition to BPO. Examples of services outsourced/outlocated include customer interaction services; the processing of credit card accounts, insurance claims and business payrolls; the creation and maintenance of information bases in the form of networked data centres and their use in the provision of information services such as help desks; and the generation of digitised records as in the case of medical transcription. Other examples of outlocation include investment in design subsidiaries and back-office facilities.
12. Figures from 'BPO: Growth all the Way, Despite the Backlash', from http://www.dqindia.com/dqtop20/2004/ArtIndseg.asp?artid=59660, accessed February 2005.
13. PTI, 'Gartner doubts viability of Indian BPOs', from http://www.rediff.com//money/2004/apr/19bpo3.htm, accessed 19 April 2004.
14. See 'Only 7–9 per cent IT jobs to move out of US', at http://www.rediff.com//money/2003/nov/19bpo2.htm, accessed July 2004.

4

ICTs for the Rural Poor: Civil Society and Cyber-Libertarian Developmentalism in India

T. T. Sreekumar

1. Introduction

The convergence of development rhetoric and information society theories in discourses on the digital divide and information capitalism marks a cyber-libertarian turn in development studies. While critiquing the tendency to conflate civil society and information society, Sparks (1994) draws attention to a widely held deterministic position that the increases in productivity brought about by information and communication technologies (ICTs) would lead to a progressive weakening of the power structures that underpin the social foundation of capitalism. Sparks highlights particularly the approach of 'new times' theorists of the now defunct magazine *Marxism Today*, of advancing the argument that communication technologies can 'undermine strong power pyramids more directly' (Sparks, 1994, p. 38). This approach largely ignores the deep contradictions of informational capitalism characterised by increasing income inequalities reinforcing development divides, social exclusion and dependency (Parayil, 2005). The new cyber-libertarian approach to ICTs and development is characterised by two interrelated arguments on the political economy of development. First, it argues that in the advanced industrialised world wider use of ICTs would create a more equitable and democratic society, and thereby obviate the need for radical social transformation. Second, it proceeds to show that the diffusion of ICTs in less-developed countries, especially in rural areas, would help to bridge the development divide. In other words, the cyber-libertarian approach regards bridging the digital divide as the most important step to emancipate the masses from poverty, because, it is believed, new technologies like ICTs would deliver rapid economic growth.

An important practical corollary of this approach has been the emergence of the notion that ICTs provide an unprecedented socio-technological possibility in developing economies for mobilising rural resources for community development, social support, employment, poverty alleviation, economic growth and increased democratic participation. This new opportunity for resource mobilisation has become the basis for civil society organisations (CSOs) to be actively involved in rural economic and social life by creating ICTs-based social enterprises to pursue these ends. Consequently, community activists, citizen groups and donor-dependent CSOs have initiated a rich variety of social experiments which, based on the use and diffusion of ICTs, particularly the Internet, claim to enhance employment opportunities; generate income streams for alleviating poverty; empower Dalits and women; strengthen neighbourhood ties; support e-governance; help overcome cultural isolation; and combat social exclusion. These ICTs-based CSOs focus their attention on the dissemination and use of ICTs in the rural sector, basing their programmes on a pragmatism of the lead role taken by donor agencies and 'social venture capital'.[1]

It appears that the rhetoric on the potential of these organisations is, more often than not, highly pitched and imbalanced. I would argue that there exists a wide chasm between the expectations and the actual benefits of CSO initiatives in rural India. While recognising the fact that the incremental values created by these experiments are important in the contexts of relative deprivation and social degeneration, we should recognise that they do not constitute a case for adopting techno-determinist models of social and economic development. Further, the model of ICT initiatives that evolved in the context of direct state interventions shows a strong reluctance to accept the role of civil society in these enterprises in any meaningful manner. Interestingly, the attempts by the CSOs to negotiate with the state for joint action in their own initiatives are also becoming increasingly problematic. Moreover, the CSOs' claims on the social benefits of the projects they undertake more often than not hinge on unsubstantiated narratives of success and achievements. These narratives end up eulogising the importance of ICTs in developing countries to justify the diversion of scarce rural resources for new ICT projects, often disregarding the opportunity cost of such programmes given the enormous need for other types of rural aid in areas such as infrastructure, health, education and agriculture.

The dominant model of these social experiments in the rural setting has been the establishment of multi-purpose kiosks catering to local-specific and clientele-based packaging and delivery of information.[2] This chapter draws upon the experience of two major projects in India, namely the M. S. Swaminathan Research Foundation (MSSRF)'s Information Village Research Project (IVRP) in Pondicherry and the Development Alternative (DA)'s TARAkendra information centres in Bhatinda, Punjab. The analysis uses the insights from recent research in community informatics to inform its framework.[3] Primary information to back up my argument was collected through fieldwork conducted during January to August 2002. I conducted extensive interviews with project officials, beneficiaries, volunteers and bureaucrats.

This chapter is divided into six sections. Section 2 is devoted to elaborating on the new prominence of the gradualist neo-liberal paradigm in development thinking and its implications for the ICTs-oriented social enterprise model as a prelude to the empirical discussions on the organisational dynamics of these initiatives. Sections 3 and 4 are devoted to understanding the formation of the so-called 'rural network society' based on the rural kiosks, which are situated in a complex social process mired in layers of tensions and contradictions, often overlooked by analysts of community informatics and social change in South Asia. In sections 5, 6 and 7, the chapter provides an analysis of the expectations and outcomes of the projects; their relations with the state and its agencies; and a critical assessment of the narratives of success which glorify superficially the achievements of these initiatives, lending credence to the cyber-libertarian development model. The chapter concludes, in section 6, with a discussion of the larger political economy implications of the deployment of ICTs for rural development.

2. ICTs and neo-liberal developmentalism: the rise of the social enterprise model

As outlined earlier, this discussion is placed against the backdrop of the emerging scenario of a seemingly unstoppable ICT fetish reinforced in the wake of glorifying narratives of information capitalism and cyber-libertarian development thinking, which has gained uncritical acceptance among policy-makers and politicians, particularly in the developing world. A vacuum has been created by the palpable dissolution of the grand narratives that defined the terms of debates in development theory and practice, such as modernisation, modes of production and dependency theories, in the last three decades of the twentieth century. This has been filled by fragmented yet powerful discourses of participation, empowerment, micro-enterprise, self-help, good governance and a host of other ambiguous interventions and concepts defined and practised by an emerging nexus of bureaucrats and CSO professionals across the Third World.[4] The issue of underdevelopment, initially viewed in terms of its geo-political dimensions, is at best seen as consisting of national, sub-regional or local-level problems that can be overcome by 'strengthening the local people', 'enhancing people's participation', 'promoting good governance' or facilitating micro-credit availability. All such 'novel' initiatives are now expected to receive a new impetus through the innovative use and diffusion of ICTs. The shifting of the terms of the debate, from the economic power of the multinationals to the 'empowerment of the marginalised', from the political struggles for freedom and democracy to the participation and self-help actions of individuals and their communities, and from the international debt crises of poor nations to the new self-help intervention of micro-credit to the rural poor, reflects the replacement of the grand narratives by a set of vaguely defined and fragmented frameworks, concepts and practices.

It might be asked how the new fragmented narratives are less clear than the older paradigms. Grand narratives such as dependency theory or modernisation theory had an idea of social transformation that was more messianic than implied by the new concepts of empowerment or governance. Modernisation theory, for example, argued that technology spillover or merely technical advances would lead to the catapulting of developing nations into the take-off stage of economic growth and, consequently, to the level of development achieved by advanced capitalist countries.[5] It was a path all economies were destined to traverse no matter how they tried to achieve it, through domestic market expansion or through internationalisation. Salvation was assured for the nation either way. Dependency theory claimed that underdevelopment was the result of internationalisation and integration of these economies into the world capitalist system, and a de-linking effected through socialist revolution would re-orient these economies into the path of true development.

The new project, however, makes no such grand claims of macro-social transformation. But this does not mean that it is devoid of any political agenda. On the other hand, it is a political intervention that, with its rhetoric, conceals its politics and economics. The hidden agenda becomes explicit only when we look at the varied modes through which the project becomes operative. The important fact about it is its fragmented character. In one region or in one locality the emphasis might be on micro-credit. In another place and time, literacy might be posited as the major concern. Emphasis may be on poverty alleviation or the empowerment of women, on another occasion. Yet, these seemingly disparate activities are part of a common agenda of development action across the Third World. However, at the same time they are thoroughly desegregated, propelled by different CSOs and funded by diverse aid agencies. The target of these operations, unlike in the case of the grand narratives, is not clearly specified. There is neither a time line for development nor a specified path. Nevertheless, that it is elusive and vague in its motives and motivation does not provide it with any immunity to criticism. Neither does the fact that no meta-theory illuminates its fragmented rhetoric obviate scrutiny.[6]

Pieterse (2000) points out two major strands of development thinking from the 1980s onwards – neo-liberalism and human development. The patterns of hegemony and theoretical explanation underlying these strands are apparently different. While the hegemonic context of neo-liberalism is the continuing prominence of finance and corporate capital, that of development is placed within the economic milieu of the performance of the newly industrialising economies and the emergence of new big markets in the Asia-Pacific region. The explanatory narrative in the case of the former is neoclassical economics and monetarism, while the latter is explained in terms of such concepts as capability, entitlement and developmental state. Interestingly, the *Human Development Reports* (HDRs) published by the UN have, curiously, tried to bridge the gap between these two strands and provided a basis for making them mutually supplementary. The HDRs do not argue against neo-liberal agenda. On the contrary, they argue for strengthening it. Discussing globalisation, HDR (1999) states:

> The challenge of globalization in the new century is *not to stop expansion of global markets*. The challenge is to find the rules and institutions for stronger governance – local, national, regional and global – *to preserve the advantages of global markets and competition*, but also to provide enough space to ensure that globalization works for people – not just for profits. (Emphasis added, UNDP, 1999, pp. 2–3)

The HDR (1999) envisages a globalisation that underscores ethics, equity, inclusion, human security, sustainability and development. While commenting on the marginalisation of poor countries in the emerging international economic regime, the report argues that 'this risk of marginalizing' does not have to be a reason for despair. Instead, it calls only for action to increase connectivity, community orientation and creativity to adapt to local needs, collaboration and finding innovative ways to fund the so-called knowledge society. The international arrangements envisaged by the report emphasise the continued role of the World Trade Organisation (WTO), the International Monetary Fund (IMF), the World Bank and a host of international NGOs and aid networks to strengthen global governance.

It can be argued in this context that this new discourse on globalisation points to the convergence of two strands of development thinking that replaced the grand narratives of development of the 1960s and 1970s. The emergence of the idea of a knowledge society has created a new field of discourse where development is decentred and moved to the margins of the debate. It appears from the thick layers of absurd reasoning that shroud this new discursive field that 'human development' means making life a little better at the local level with the help of CSO professionalism and a dose of 'Development'. The capital D is intended to showcase the unbridled expansion of markets, which supposedly propels economic growth. In other words, globalisation, as it is understood in relation to this new development thinking, is economic globalisation,[7] manifested by the reinforcement of market forces.

Economic globalisation is thus an integral part of the expansion of the knowledge society across the world. United Nations (1999, p. 138) defines economic globalisation as growing economic interdependence of countries worldwide marked by an intensification of the international division of labour due to phenomenal growth in international foreign direct investment flows. As a result of this, it is envisaged that there would be massive increase in the volume and variety of cross-border transactions in goods and services, capital mobility and international migration, and more rapid diffusion of new technologies. Knowledge is identified as the key factor of production, surpassing land, labour and capital.

The increasing visibility of the information economy is primarily manifested by the proliferation of 'technopoles', which are situated in the United States, England, continental Europe, and newly industrialised countries (NICs) in South-East Asia and Japan. Castells and Hall (1994) argue that technopoles exemplify the reality that cities and regions are increasingly being structurally reconfigured. They are also conditioned in their growth dynamics by the interaction of major global historical processes. These include a technological revolution based on information technologies (including genetic engineering), the formation of a global economy that works as a unit in a worldwide space. This change is characterised by new forms of capital (financial), management, labour, technology, production arrangements and markets, where horizontal networks replace vertical bureaucracies and flexible production arrangements replace standardised mass production.[8]

United Nations (1999) also argues that rapid advances in ICTs promote growth by increasing efficiency, integrating markets and supporting growth in services. Nevertheless, it is also true that the benefits of globalisation are spread highly unevenly across nations. There is a widespread belief that international technology spillovers are becoming increasingly transparent and smoother with the new phase of globalisation and liberalisation of domestic economies in developing countries. But in reality, when the process of globalisation set off in a race to claim new knowledge, the global gap between haves and have-nots has, in fact, widened (UNDP, 1999). This could be due to a re-enforcement of the patterns of global economic processes, which historically and politically hindered the development of low-income countries. Also, the more deep-rooted structural problems of domestic economies interacting with the rapidly evolving international finance and technology markets exacerbated the situation. The strategic significance of these aspects is underscored by the differential progress made by low-income countries in reaping the benefits of global information economy expansion.

The strong push made by many regions to become 'the next Silicon Valley' has mostly failed. Castells and Hall (1994) argue that the magic formula often worked out by opportunistic consultants is a small dose of venture capital, a university or technology institute and fiscal and institutional incentives to attract high technology firms. The magic formula, however, often wrapped in a glossy brochure and futuristic name, does not help build 'the next Silicon Valley'. According to Castells and Hall, 'the world is now littered with the ruins of all too many such dreams that have failed or have yielded meagre results at far too high costs' (p. 8).

If the 'technopole' model has failed to deliver in the developing countries except in a few places such as Bangalore,[9] in many countries, particularly in South Asia, new 'social venture capitalists' are now founding organisations and launching projects that claim to generate scientific synergy and promote rural economic regeneration through the application of information technology in rural and semi-urban localities. These programmes and projects have been internationally acclaimed as making a tremendous contribution towards bridging the digital divide as well as providing enormous opportunities for reorganising rural economy by generating new income streams and employment for the rural poor.[10]

The cases discussed in this chapter, the IVRP and TARAkendras, are widely acclaimed examples of organisations that claim to have delivered these promises based on the social enterprise models. The goals and achievements of these two organisations and their influence in bridging the digital divide as well as the development divide in rural India will be scrutinised below. Further, the question of whether the newly emerging rural network society is more equitable and democratic will also be examined.

3. The Information Village Research Project

The IVRP, initiated in 1998, aims to provide sustainable food security in rural areas of Pondicherry, in South India. The International Development Research Centre (IDRC), Canada, supports the project.[11] The concept is based on the needs of farmers for information on sustainable agricultural practices, credit and marketing of produce. The IVRP centres were originally designed to provide information and technical assistance to farmers through packages prepared locally making use of both indigenous knowledge and modern science. The project aims to add value to villagers' generic information to render it more location- and context-specific by using ICTs. Nevertheless, the project has set for itself a wider range of objectives. First, it plans to set up village information kiosks that enable rural families to access a basket of modern ICTs. Second, these kiosks are also used to train educated youths, especially women, in rural areas to operate the information centres. Training is aimed to empower the youths in the organisation and maintenance of a system that generates locally relevant information from generic information. Third, the project aims to take up maintenance, updating and dissemination of information on entitlements to rural families by providing access to various governmental schemes and programmes from departmental websites and other governmental sources. The fourth objective is related to the action research priorities of the project, such as conducting impact assessment based on surveys, participatory rural appraisal and so on. Finally, the project aims to show that it serves as a replicable and sustainable model in information dissemination and exchange in rural areas that use advanced ICTs. Various aspects of the evolution and sustainability of the project will be discussed below.

The project was launched after conducting a detailed village level survey to assess the communication habits and channels in the rural areas, especially among the poorer households. It is claimed that due attention was given at the time of the inception of the project to make it pro-poor, gender-sensitive and environment-friendly.

The formation of information centres is a complex social process, encompassing a wide spectrum of activities spanning several months. As mentioned earlier, initially a survey is conducted to understand the 'mindset' of the people. As pointed out by a middle-level functionary of the project: 'We wanted to see if the community is willing to take the responsibility for providing space, voluntary service and electricity.'[12] The technical personnel of the project were aware of the usual scepticism of the rural people towards programmes initiated by CSOs. This happened when the kiosks were initiated despite the fact that the MSSRF had been working in Pondicherry since 1995 promoting their BIO-Village Project, which is claimed to be a comprehensive attempt to integrate all aspects of village life, including its environment, in formulating micro-level planning.[13] After the first survey, several meetings were held with the villagers using rural participatory appraisal techniques (Balaji et al., 2001). Organising such meetings in itself is a difficult task given the local political conditions. According to a respondent, 'There is an added difficulty in Pondicherry due to the virtual absence of any formalised Panchayati Raj Institutions. The local government is called *Nattaimai*,[14] which is the rule of the traditional village Headman. Unless everything is channelled through them, no project can take off.'[15]

It is with the permission of the *Nattaimai* that project co-ordinators have to liaise with youth associations, women's associations and the Village Development Councils in order to create a friendly environment for holding the meetings, which aim to communicate to the villagers the benefits of establishing an information centre in their locality. In most places, the *Nattaimai* have been very co-operative. Resistance, wherever it occurred, was not due to any inherent opposition to the project's modernisation plank. As Kankeyan, head of the *Nattaimai*, puts it:

> Feeble but serious initial resistance to the project was based on two major interrelated problems. First was the demand for free resources such as space, volunteer services and electricity by MSSRF. This problem was actually an offshoot of another problem i.e., neither the MSSRF staff nor their local supporters were able to tell anything concrete about the benefits that the projects can bring.[16]

This is corroborated by the reflections of the project staff. They remember that during the initial period before the first centre was set up, the going was extremely tough.

It is claimed that the content of the programme also evolved through a process of continued dialogue with the community. The villagers themselves identified volunteers in most cases. The MSSRF had stipulated that the basic qualification of the volunteer should be between five and eleven years of schooling. But, overwhelmingly, the volunteers are technically qualified youth, except in a few places like *Embalam*.[17]

3.1 Local elites and the management of kiosks

It is important to understand that the process of launching the kiosks and their subsequent management is deeply enmeshed in the social realities enveloping the implementation of such initiatives in rural India. The participation of the local civil society is marginal or notional, while village elites and temple trustees acquire more responsibility for the day-to-day administration of the kiosks. The temple trustees in Indian villages are mostly upper-caste Hindus. Lower-caste Hindus and Dalits are often not allowed to interact with them on an equal footing in many places, even though untouchability and unseeability are illegal. Even today, in many places Dalits are prevented from drawing water from public wells and ponds in many rural areas in India, and atrocities against Dalits and Adivasis (aborigines) are reported frequently in the media. It appears that Pondicherry is no exception. This is evident from the fact that the MSSRF itself has been forced to close down some of its kiosks where Dalits were not allowed to enter. A sympathetic report on the project (Krishnakumar, 2001) confirms this. Since the MSSRF cannot change the social setting, they shut down the kiosk if Dalits are not allowed to use the facilities. According to the report, the MSSRF closed two centres for this reason.

The very fact that two kiosks had to be shut down in a small and relatively developed state like Pondicherry due to the practice of untouchability is significant in understanding the local power dynamics of rural India. The village headmen and temple trustees have an upper hand in the management of the kiosks. Even though some of them are not explicitly practising untouchability, the impact of this influence on participation from and acceptance by Dalits and non-Hindus could be problematic given the deep divisions along caste and religious lines that still exist in rural areas at a time when secular institutions in India are increasingly penetrated by caste forces hostile to non-Hindus and Dalits.

The complex nature of the formation of information kiosks in Pondicherry points to the role social and economic factors play in the shaping of the project. Most importantly, it is worth noting that the consent of the local elite, availability of unpaid labour and free availability of space and electricity (provided by the community either through the income of the temples controlled by upper castes or revenue from local bodies controlled by the political elite) are some of the critical inputs required to launch the centres.

4. The TARAkendras

The TARAkendras in Bundelkhand (Uttar Pradesh) and Bhatinda (Punjab) were launched by TARAhaat, an Internet portal that supports a network of franchised computer kiosks. It aims to deliver a wide spectrum of services through ICTs and expects to earn revenues by pricing these services, as well as by levying membership fees and commissions. One of the striking features of TARAhaat is that it was not conceived as a voluntary venture from the very beginning, although its sponsoring agency (DA) is a CSO operating in Bundelkhand. The concept of TARAhaat took shape in 1998, although it became functional only in 2000. The delay was due to the formidable challenges that CSOs face in executing technology-based projects in rural areas. The first few kiosks were established in Bundelkhand, a region where the DA has been active for several years.

The DA was founded in 1983 with the objective of developing models for sustainable livelihood for the rural poor. The formulation as well as the implementation of the concept of TARAkendra materialised as a result of a group of young professionals committed to the cause of social development coming together in the region.[18] It is a for-profit venture, and pays taxes.[19] Nevertheless, the progenitors and functionaries of the organisation do not attach much significance to its profit orientation other than for demonstrating that such market-based ventures from civil society foundations are commercially viable.[20] It is a business model with a relatively stronger emphasis on the social economy of the region (Peterson et al., 2001). Moreover, the organisers hope that TARAhaat's NGO roots and an ownership structure that allocates a 51 per cent share to the non-profit Sustainable Livelihoods Foundation will ensure that social objectives are not overlooked (Peterson et al., 2001, p. 4).[21]

The TARAhaat follows a 'business model' different from the MSSRF model, which depends on voluntary unpaid labour and free access to scarce rural resources. Nevertheless, the process of setting up a TARAkendra is quite similar to the processes and patterns that underlie the launching of kiosks by non-profit initiatives like the MSSRF. Identifying *Mandis* (small towns or villages) in Bhatinda where TARAkendras can be set up was initially undertaken in collaboration with the District Rural

Development Agencies (DRDA). But this alliance ran into problems following the transfer of a bureaucrat sympathetic to the concept of rural kiosks. Despite the lack of support from the DRDA, the TARAhaat expanded its operations. However, the process is quite complicated and requires constant interaction with a wide spectrum of individuals and institutions in the small towns or villages. The TARAhaat sends its business managers to tackle the initial problems related to the setting up of a rural kiosk. Within the organisation this activity is identified as business promotion.[22] The work involves winning the confidence of the local elites and then finding an entrepreneur willing to collaborate with the TARAhaat on its terms of franchising. The process is well summarised by a TARAhaat official:

> When I go to a *Mandi* I first try to meet the *Aarthyas,*[23] who are the most reputed people in many villages. Talking to them and winning their confidence is essential to broach the concept of TARAhaat with local community. I also talk to the teachers and other professionals in the village. Sometimes it is also necessary to talk to people in the neighbouring villages also. It takes time for them to understand the concept and respond positively. Several visits are required to put across the idea.[24]

The process, thus, involves an initial contact with the village elites identified by the business promoters and subsequent discussions with them as well as other villagers. An entrepreneur with a minimum investment capacity has also to be identified and given training in the mode of operation of the organisation as well as in running a kiosk. The paid staff of the kiosk, generally computer-educated youth, also has to be trained. A TARA Academy has been set up with the objective of serving this requirement. Once the centres are established, the TARAhaat attempts to consolidate the operations by providing a range of community activities and programmes, including seminars and group meetings, through carefully structured institutional means.

5. IVRP and TARAkendra: a comparison

Both the IVRP and TARAkendras are organisational innovations in initiating ICT-based projects for social and economic transformation. The long- and short-term objectives set by these organisations reveal a high level of sensitivity towards the contexts within which rural deprivation and economic degeneration occur. I will concentrate on three major aspects of their performance as social enterprises and seek to study the dynamics of success and failure as they arise in different social settings. The first set of issues that is being addressed relates to the expectations and actual achievements of these initiatives. Although they are widely separated in space, these initiatives have a set of objectives with common tangents, in that they bear similarities in terms of their larger social content that make a comparison possible and relevant. The second set of issues we address pertains to the contested nature of the state–civil society relationship emerging from the developmental trajectories of the IVRP and TARAkendras. Reference to the state is essential in conceptualising the social economy, even though the extent of dependence on the state and the nature of conflicts and co-operation between the CSOs and the state vary widely. Finally, we will look at the sustainability question and discuss the narratives of success that are often disseminated through the media as indicators of growing demand for IT-enabled services in rural India. These media reports reflect on the viability and sustainability of the projects to legitimate the implied cyber-libertarian development paradigm.

5.1 Expectations and outcomes

While the benefits of the spread of ICTs in rural areas have considerable social and economic impacts, the projects cannot deliver the range of outputs that were initially envisaged due to a multitude of factors, such as limited local participation; lack of availability of local resources; fractured relationships with the state and its agencies; limited integration with other local CSOs; and the exogenous social and economic environment that these enterprises are unable to control. These factors are often overlooked when projecting the promises of the projects. This approach has consistently led to the overestimation of expected outcomes while underplaying the pitfalls.

In the case of the MSSRF, it can be seen that it relies heavily on local resources for office space, non-paid labour and electricity. Facilities at the centres are highly uneven because most communities are poor and socially backward. They are unable to acquire the infrastructure required for setting up the kiosks as well as maintaining them (see Table 4.1). The volunteers invariably pointed to either indifference or inability of the local body as the reason for lack of facilities in their kiosks. The Poornankuppam and Thirukanchipet kiosks are good examples to illustrate the ways in which limitations of resources and low demand for ICTs-enabled services could stand in the way of the dreamy projections of the successes of such ventures in rural India.

Table 4.1 Unevenness in basic infrastructure facilities at selected IVRP centres

	Veerampattinam	Embalam	Poornankuppam	Thirukanchipet
Connectivity	Phone connection	Phone connection	Not connected	Not connected
Internet	Available	Available	Not accessible	Not accessible
Web camera	Available	Available	Not available	Not available
No. of PCs	8	3	3	3
Speaker system	Available	Not installed	Not installed	Not installed
Allowance for volunteers	Rs 500 paid by Panchayat*	No allowance	No allowance	No allowance
Spread spectrum	Installed	Installed	Not installed	Not installed
Solar panel	Not installed	Installed	Not installed	Not installed
Siren	Not installed	Installed	Not installed	Not installed
Wireless phone	Connected	Connected	Connected	Connected

Note: *Allowance was being paid only to one of the volunteers and the payment had become irregular at the time of my investigation
Source: Based on author's fieldwork

5.2 The Poornankuppam centre: an evaluation

A visitor to the centre can instantly note the presence of an active band of self-help groups in the village by looking at the display boards placed at the main junction. Several women self-help groups functioning in the village bear such names as Mahatma Gandhi, Jhansi Rani, SEVA and Sarojini. But none of these organisations is involved in the management of the kiosk. One of the four volunteers in charge is a 23-year-old youth who has completed his graduate diploma in Computer Applications. He joined as a volunteer in August 2001 when the centre was on the verge of being closed down for want of personnel to manage it. He remembers: 'I first came here as a user and met the staff of MSSRF. After my graduation I volunteered along with three of my friends when the centre had been closed for more than two months for lack of volunteers to run the centre. I went to the Vilanur office and offered my services with my three friends.'[25] One of the reasons, perhaps, for the reluctance of the local youth to associate with the centre is its relative lack of infrastructure.

The temple offers the space and electricity for the centre free of charge. The relationship with the temple trust is relatively tension-free except for during the festival season, when the trust asks the centre to vacate the space for opening the reception committee office for the festival. But through lobbying of the Panchayath (local body) members who are sympathetic to the centre, the volunteers always succeed in revoking this decision by the temple trust. However, the relationship of the centre with the Panchayath appears to be very weak, with the Panchayath offering no help, support or assistance for the project. It appears that the kiosk cannot afford to pay the monthly rental charges of the phone, a paltry sum of Rs 250. According to my informant, there were very few foreign visitors and little media exposure to this centre, unlike the one in Veerampattinam.

The fax message from the Vilanur centre reaches the centre on weekdays around 10.30 a.m. and includes: (1) Price list, (2) Weather Report, and (3) Daily News. This is immediately placed on the notice board. On average, a maximum of six to ten farmers would visit the centre to make use of this information. The volunteer is worried that their numbers would be still less if the centre decides to charge for the information in the future to make it a business model. The user fees that the centre gets from the services are just nominal. It is in the month of May during the school vacation that the centre gets the largest number of students to join the computer course. The register for May 2002 showed that, on average, seven to eight students visit the centre for the purpose of learning computer operating skills. It appears that the centre attracts 10 to 15 students for attending its computer courses by paying Rs 50 as monthly fee. Some people also use the centre as a telephone call booth, since it provides Internet phonecalls for Rs 2. Nevertheless, the revenue from the centre cannot be considered as breaking even if subsidies are withdrawn.

5.3 The Thirukanchipet centre

This centre represents a classic example of how resource-poor lower-caste Dalit villages struggle to maintain the kiosks without subsidies. The Panchayat provides space and electricity for the centre by default, since the centre operates in a small room where the Panchayat TV is kept for villagers to watch programmes aired on the government channels. Before the TV arrived, this room was used for listening to the radio. While one can see a gradual progress of the village centre from radio to television and eventually to a computer, the inability of the Panchayat to provide additional resources for running the kiosk has left the volunteers to depend solely on the subsidy given by the MSSRF. Although there is a Dalit temple in the village, the centre does not get any assistance from the temple management. One of the volunteers of the centre explains:

> There is a temple nearby. The temple management is not able to offer us any help, because being a Dalit temple they themselves have very little resources. The Panchayat has not agreed to pay the phone bill if we take Internet connection. Since the people are Dalits and poor the local Panchayat is also not rich. So we have no Internet, no phone connection, no spectrum, no siren and no speaker system.[28]

However, he hopes the Panchayat will soon find a solution to the problem of phone connection and will somehow find the means to pay the utility bills. The user fee that the centre generates is negligible. Children playing computer games are charged at Rs 4 per hour. But not many parents in the village can afford this sum. Some schoolchildren, however, get an opportunity to learn some of the basic computer skills at the low user fee. According to the informant, farmers are not very enthusiastic about the centre:

> We campaign and canvass people to use the facilities in the centre. I would say that students and the youth of the locality mainly use the centre. The MSSRF has said that they will withdraw all subsidies in 2004. By then we are supposed to stand on our own.[27]

The experience of these two MSSRF kiosks points to the resource constraints that cripple the knowledge centres. Consequently, the kiosks are established only in a few villages and even in the case of centres with more facilities they are mostly visited by children for playing computer games and sometimes working on their academic projects. In some centres the value-addition to generic information and its dissemination appears to be useful. But in what ways this information becomes crucial in everyday life depends on contexts and it is difficult, if not impossible, to assess how it contributes to productivity improvements.

In the case of the TARAkendras, the success of each enterprise depends upon the market size of the *Mandis* where it operates. The TARAkendras were expected to expand and take root at a faster pace than what actually materialised. An early projection on the possible proliferation of TARAkendras estimated phenomenal growth in the number of kiosks as well as users by the year 2005 (Khanna, 2001). In reality, the progression has stagnated and TARAhaat has closed down its operation in Bundelkhand. The euphoric expectations have been totally unrealised. As discussed earlier, the real difficulties in setting up a centre are formidable, and it is unrealistic to make overly optimistic growth projections.[28] The TARAkendras were hoping to get 10 to 12 or more students per computer. Envisaging a commercial model of computer education offering different kinds of content delivery, it was hoped that they would 'get about 12 registered students per computer at any time in a TARAkendra' (Prakash, 2001).

The performance of different centres, as in the case of IVRP kiosks, depends heavily on local-specific factors. The education model, however, has failed to appeal, as the experiences of several TARAkendras illustrate. Massive job creation and poverty alleviation are lofty goals, but extremely difficult to pursue, with an overemphasis on the potential of new technologies. This is evident from the fact some of the TARAkendras with a longer history have either been closed down or disfranchised. Moreover, the older centres are performing badly in terms of student enrolment and user turnout, making them economically unviable. The experiences of two TARAkendras, in Gonniana and Lehra Mohabbat, illustrate this point well.

5.4 The Gonniana TARAkendra

This centre was established in April 2002 with an initial investment of Rs 200,000. The franchisee of this centre holds a postgraduate Diploma in Management from the UK and a Business Administration Diploma from Hyderabad, India. He opted for a TARAhaat franchise for two reasons: the belief that the public would value a TARAhaat certificate in computer education since DA (its sponsor) is recognised by the Ministry of Science and Technology, New Delhi, and the promise that TARAhaat would make available the curriculum in the local language. He also felt that the fee structure is competitive.

Some private-sector outfits in the locality attempted to compete by lowering their fees, but could not sustain the price war. The franchisee has purchased a generator to offset frequent power failure during the day. Including debt servicing, rent and other payments, he incurs a monthly expenditure of Rs 10,000 to run the centre. This is inclusive of the salary he pays to the trainers. He has 26 students and earns between Rs 12,000 to 15,000 per month. He considers this breaking even, and expects that the centre will become viable when the number of students increases. He has taken a keen interest in the community-based activities of the TARAhaat and has taken steps to launch the innovative programme for women and children called TARAbaathcheeth in his centre. He believes that the community-based approach of the TARAhaat will help his centre to attract more students as well as clients for the services that he can offer through TARAhaat.com. He also runs the new Practical English Course designed by TARAgyan, for which he receives an enthusiastic response from students.

5.5 The Lehra Mohabbat TARAkendra

Unlike the kiosk in Gonniana, the Lehra Mohabbat centre established in March 2001 is one of the older kiosks. The franchisee pooled the initial investment of about Rs 280,000 as a loan by pledging one acre of land and pays Rs 5000 per month for debt servicing. He incurs a monthly expenditure of Rs 3000 towards telephone and electricity charges and pays a monthly rent of Rs 4500. A salary of Rs 2500 has to be paid to the trainer. His total monthly earning from the TARAgyan courses is a meagre Rs 6000 and an additional income from printing and Internet services is Rs 1000 to 1500. He runs a deficit of Rs 6500 to 7000 per month, which is compensated by family income and other sources. The student enrolment has come down from 40 in the first year to 12 in the second year, causing a marked drop in his earnings.

The experiences of both the IVRP and the TARAhaat suggest that the success of their social enterprises depends heavily on several factors, which are contextual as well as structural. The CSOs are excessively credited with an overwhelming record of creating jobs, enhancing capabilities, generating income streams for the poor, developing locally relevant services and making markets work for the underprivileged. The jury of the Stockholm Challenge Award, for example, quoted in MSSRF (2002a), says:

> Today, thanks to Information Village Research, ten villages near [*sic*] Pondicherry, India, are linked with computers, providing information on such aspects as health, crops, weather and fishing conditions. These new technology tools are bridging the economic and social divide between the haves and have-nots. They are empowering everyone with knowledge and opportunity by an inclusive use of local languages and a multimedia format that allows all to participate.

Similarly, the TARAhaat was initially seen as 'an impressive blueprint for transforming the poverty-riddled country side into a vibrant economy. It hopes to catapult the villages from their 19th century existence directly into the digital era by creating jobs, promoting sustainable livelihoods and transforming rural marketing systems – things that will lead to a quantum jump in the pace of rural development' (Jishnu, et al., 2001).

While such tall claims are a far cry from reality, the projects have been marginally successful in introducing ICTs into the rural setting. Understanding the pitfalls and demystifying the projects are essential for creating a balanced perspective on the incremental benefits of harnessing ICTs for improvising developmental strategies in rural India. This is particularly important, since the actual possibility of mobilising additional resources required for sustaining these projects if the sponsor CSO withdraws subsidies is highly unlikely, in contrast to the overly optimistic beliefs held by some CSO functionaries.

6. State–CSO relations: the emerging contradictions

Interventions in the social economy by ICTs-oriented CSOs are highly dependent upon continued support from the state and its agencies, since most of the services that these enterprises offer in the villages have historically been the responsibility of the government and the major chunk of the information demanded by the villagers concerns government services and welfare schemes. Nevertheless, the relationship between the CSOs and state agencies in this sector is highly precarious and uneven.[29] In its initial efforts to anchor in Pondicherry, the MSSRF sought immense logistical support from the government, particularly from the Department of Agriculture.[30] However, the rapport did not extend to a joint operation.[31] On the contrary, their efforts began to diverge, culminating in the emergence of parallel IT centres called Uzhavar Uthavi Agam (UUA, literally Farmers' Support Kiosks) set up by the Department of Agriculture (2002). The objectives of the UUA are strikingly similar to the MSSRF centres, yet the Department of Agriculture does not want any formal or informal link with the MSSRF. Still, the MSSRF functionaries are optimistic about linking up with the government agencies.

The MSSRF continues to maintain links with the Department of Science and Technology in New Delhi. But even the functionaries of the MSSRF would consider this a notional tie. The nature of relations is limited to the provision of one-third of the Rs 250,000 required for setting up a new kiosk. Another wing of the government, which has some ties with the MSSRF, is the Department of Rural Development. This department has suggested that out of the 500 micro-finance units run by women under DRDA, they are willing to identify a few best units to run information kiosks. The required amount of Rs 250,000 would be provided by the department as loans with a 30 per cent subsidy. The absence of strong linkages with the government agencies severely limits the MSSRF's ability to provide e-governance services to its clientele.

The TARAkendras face a similar problem, though the situation is somewhat different from the MSSRF's experience. The failure of TARA-haat to deliver on its promises of e-governance has dented its credibility with the franchisees and customers (Kaushik and Singh, 2004). This happened partly due to the discontinuity in TARAhaat's expansion. Initially, it could link more effectively with the DRDA and the prospects of delivering e-governance-related services looked relatively bright. The break-up with the DRDA incapacitated the TARAhaat from accessing government information and providing e-governance services. However, the progenitors of the TARAhaat concept attempt to underplay the impact of this failure.

The experiences of both the MSSRF and TARAhaat show that the social economy initiatives do not always receive state support and patronage. The relationship is more often than not fractured and precarious. One of the reasons for the state's indifference to the CSOs is based on the premise that experiments sponsored by the state and implemented by its bureaucrats would work better and deliver more efficiently than those run by the CSOs. Further, such entities, often called QUANGOs (Quasi-Autonomous Non Governmental Organisations) can be more systematically controlled and channelled by the state.[32] The strategy of the state seems to be twofold: to keep its involvement in CSO activities to the minimum and to limit the representation of CSOs and other civil society actors in its own initiatives, including the functioning of the QUANGOs. Further, the possibility of establishing easy legitimacy for QUANGOs-based developmental work with international donor agencies has led the state to pursue its own strategies, thus sidelining the CSOs almost completely.

7. Narratives of success and the sustainability puzzle

Despite the problems, there are several successes credited to these projects. These successes are, more often than not, communicated by the organisations as anecdotes and stories, which primarily narrate individual or group gains as a result of the introduction of the project. A closer look at these stories, however, provides a rather different picture of the nature of the rural network society that these projects aim to create. Illustratively, some such stories about the MSSRF project can be scrutinised to analyse the conditions for the real success of ICTs-based social enterprises, by juxtaposing the narratives against grounded realities. Surprisingly, even the evaluation of the performance of the centres is sometimes carried out on the basis of these stories, as admitted by a report by the PANAsia Telecenter Learning and Evaluation Group of the IDRC which conducted the evaluation of the MSSRF kiosks. According to this report:

> Prior to the mission, it was decided that the group would focus on stories from telecenter users and operators. Stories, whilst anecdotal, offer a rich picture of the impact of ICT interventions in local, complex and dynamic social settings. They are accessible and verifiable during short visits and they acknowledge the often indirect influence that development interventions have on their beneficiaries. (IDRC, 1999)

A frequently told story is about an old woman and her cow. This story has at least two versions. In the first version Bubrayan Panjali, 'a round-faced illiterate woman' from Kizhur village in Pondicherry, was able to get a veterinary doctor to attend to her cow, her 'only source of income', which was in labour and would have died without the information from the MSSRF e-kiosk. One report says that 'for five days the cow moaned while in labour' (Dugger, 2000). The report continues:

> Word of Mrs. Panjali's woebegone cow soon reached Govindaswami, a public spirited farmer who uses one name. The village computer, obtained through the Swaminathan Foundation, is in the anteroom of his home. The computer is operated full time and for no pay by his 23 year old college educated daughter, Azhalarasi, who used it to call up a list of area veterinarians. One doctor arrived that night and . . . then dragged the calf into the world.

In the second version of the story, it was Panchavarnam's pregnant cow that had been in labour for four days but could not deliver her calf. For Panchavarnam the cow's survival was crucial as it was her 'only source of income' after her husband's death. News of the cow's plight spread, and G. Ezhilarasi, a college student who operates a computer from an ante-room in her house, surfed the Internet for veterinarians and contacted several of them in the area. On the fourth day, one doctor responded to the message. He came to the village and assisted in the delivery (Krishnakumar, 2001).

Setting aside the discrepancies in the stories and the details that change with re-iteration, it is still evident that the story has been blown up out of proportion in a bid to show that the impact of ICTs is enormous. During my fieldwork I found that it is true that there are no veterinary doctors or hospitals in Kizhur. Nonetheless, the nearest one is only just three to four kilometres away in the village of Sivarandhagam. The hospital works every weekday as well as on Saturdays and Sundays. The hospital has been functioning since June 1998 and its phone numbers are listed in the local telephone directory. If no one turned up to help the old woman for four or five days it shows the breakdown of the offline support systems in the neighbourhood. And, more interestingly, none of the versions of the story claims that the old woman approached the e-kiosk for help. The story is thus a typical example of crafty narration transforming ordinary everyday events into parables to show rural change by means of computers and the Internet.

Just as much has been reported about the weather information provided to the fishers in Pondicherry through the information kiosks. Actually, this facility is available only in one kiosk in Veerampattinam. It is claimed that weather information about storms and so on downloaded from the public website of the US Navy and conveyed to the fishers in Pondicherry helps them to save their lives (Kapoor, 2001). Thermal wave maps provided by the kiosks supposedly enable the fishermen to snare bigger catches (Jishnu et al., 2001). It is reported that 'fishermen in Pondicherry do not go out to sea without first looking up the thermal wave charts and weather bulletins provided by a local network' (Jishnu et al., 2001, p. 30). Interaction with local fishers revealed that the weather information provided to them is not that crucial.

There are two aspects to this observation. The fishermen's traditional methods of weather predictions based on cloud formations and wind directions are more or less accurate, according to the local people. The reason why some fishermen still venture out to sea in rough weather is due to their acute poverty. When asked about the importance of weather information, one of the fishermen in Veerampattinam said: 'We know what will happen. Still we go because we don't want our children to starve.' Caught between the notion of 'tradition' and 'modernity', the village head was, however, apologetic. He said, 'The weather information provided by the centre is credible.' The question of local knowledge versus modern science discussed in the growing literature on post-development appears to be relevant here.[33]

These stories fall apart as credible indicators of the social and economic benefits of the projects as we probe further the contexts and backdrops of their origins and trace them through the layers of iteration which turn them into new folklores of the rural network society in the making. The question of sustainability, hence, has to be analysed on the basis of more solid criteria than what these stories could teach, even in cases where there is a grain of truth in them. The potential for sustainability of the rural kiosks depends heavily on local contexts, which the projects are often unable to control. Keeping costs low depends on several factors. However, for the MSSRF centres the equipment is provided by donors. For the TARAkendras, the cost is borne by the local entrepreneurs who hope to generate sufficient income and gain a profit. Recurrent costs of Internet access, office stationery, electricity, maintenance and repair, telephone charges and salaries, if any, have to be met by the revenues and any other subsidy received. The revenue sources for these projects are the user fees they hope to receive from the poor rural users.

User fees, the expected source of income to sustain the centres, are difficult to generate. This is especially the case because any agglomeration effect expected through network effect (Metcalf Law) was not forthcoming due to poor network expansion. The Metcalf Law states that the value of any complete network such as the Internet would tend to grow at the rate of the square of the number of users, as opposed to a simple linear growth. In other words, the law states that the value realisation from networks is critically affected if the network is quite small. Our analysis of the MSSRF's IVRP kiosks and the DA's TARAkendras shows that the potential for enhancing the network is severely limited and these projects have stumbled upon the challenges of expansion due to a variety of factors. In Pondicherry, for example, the progenitor of the project had hoped that the government would come forward to facilitate kiosk proliferation. However, in most projects the relationship between the state and its agencies has been fractured or ambivalent concerning benefit from partnerships. The cyber-libertarian approach of gradualist development mediated through ICTs grossly ignores such limitations faced by these projects.

8. Discussion and conclusions

Cyber-libertarian development thinking has pervaded contemporary discussions on development. The intervention of ICTs-oriented CSOs in development initiatives is often eulogised and receives undue thrust in the media as well as in international development discourse. Robert Wade (2002) has rightly identified the common features of this approach, which he calls 'groupthink'. The groupthink, according to

Wade, hinges on some key assumptions about the emerging global reality of ICT diffusion. Wade points out that groupthink maintains the position that the digital divide is the major unequalising force in the contemporary world economy, and could be addressed by supplying more ICTs to developing countries. Groupthink also views ICTs as an 'inherently enabling meta-technology that can bypass or leapfrog institutional and infrastructural obstacles' (Wade, 2002). This would in turn imply the fallacious position that the standard cost–benefit comparisons are irrelevant in the case of ICT projects, since ICTs form a new social and economic development paradigm.[34]

A scrutiny of the performance of two CSOs, the MSSRF's Information Village Research Centres (Pondicherry) and the DA's TARAkendras (Bhatinda), shows a wide chasm between the expectations and actual benefits of CSO initiatives in rural India. Contrary to popular belief, these social enterprises are not rooted in the resources of the local economy. Moreover, their prospects of evolving into developmental models capable of drawing on local community resources for sustenance appear bleak. Their ability to contribute to local economic regeneration, through such claims as job creation, developing local services and markets, enhancing local skills to inculcate entrepreneurship and building social capital, is yet to be materialised.

CSO initiatives are not universally replicable models for configuring social enterprise projects based on ICTs. The incremental values created by such experiments in India's rural setting are nevertheless important given the extreme deprivation and social degeneration in the rural areas. Nonetheless, they do not constitute a case for adopting techno-determinist models of social development.

State–civil society relations in ICT projects are marred by tensions and contradictions. The state-led projects have tacitly followed a model of limiting the role of civil society, while the social enterprise models of CSOs show varying degrees of failure in making the state respond to their signals.

There are, indeed, larger lessons to be learned from the brief history of the Internet in rural India. The social enterprise models depend heavily on direct and indirect subsidies and are effectively controlled by the local elites. This tends to reinforce, rather than bridge, the existing social divides in the villages. The economic sustainability of the projects, on the other hand, rests on the potential for enhancing the rural network society to reap the benefits of economies of scale and the network effect. It is evident that the deployment of ICTs in resource-poor regions and nations cannot be based on the over-hyped narratives of rural e-topia now pervading the discourses on new technologies and their potential. The CSOs initiating such programmes and projects have to understand the importance of assessing the complexities of the realm of their social action. As Castells (2000, p. 76) puts it, 'the complex matrix of interaction between the technological forces unleashed by our species' is a matter of inquiry rather than of fate.

Notes

1. Social venture capitalists are individuals and CSO leaders who are willing to make the necessary social and financial investment in profit-oriented social economy projects. Satish Jha, CMD, James Martin & Co, who has invested in TARAhaat and Drishtee, described himself as a social venture capitalist. Interview, 6 August 2002, New Delhi.
2. For a review of the growth of rural information kiosks in India, see Sood (2003).
3. For a recent collection of articles on various aspects of community informatics, see Keeble and Loader (2001).
4. Petras (1999b) observes that the shift involves an attempt to divert attention from the sources and solutions of poverty. For him, 'To speak of microenterprises instead of the exploitation of overseas banks as a solution to poverty is based on the false notion that the problem is one of individual initiative rather than the transfer of income overseas' (Petras, 1999b, pp. 434–5).
5. Rostow's stage theory (Rostow, 1960) has been a favourite standard explanation of this transition within the modernisation paradigm. The stage theory frames societal evolution from a backward traditional agrarian economy to an industrial society stage of high mass consumption.
6. A new philosophy of gradual betterment of civic life in poor countries is projected as the underlying rationale of this global programme executed locally but planned globally. Many CSO professionals see a vindication of their work in this new philosophy. Muhammad Yunus, founder of the Grameen Bank in Bangladesh, aptly captures this mood: 'When I was arguing that helping a one meal family to become a two meal family, enabling a woman without a change of clothing to afford to buy a second piece of clothing, is a development miracle, I was ridiculed. That is not development, I was reminded sternly. Development is growth of the economy, they said; growth will bring everything. We carried out our work as if we were engaged in some very undesirable activities. When UNDP's *Human Development Report* (HDR) came out we felt vindicated. We were no longer back-street operators, we felt we were in the mainstream.' (Yunus quoted in UNDP, 1999, p. 15)
7. The debate on globalisation has produced a voluminous literature. Nevertheless, the critical points of contention revolve around four fundamental issues pertaining to the extent, nature, constraint and impacts of globalisation (Held et al., 1999). There are also authors such as Petras (1999a) and Sutcliffe (1999) who believe that imperialism is, analytically as well as heuristically, a better word to capture the essence of globalisation.

8. Harvey (1989) attempts to capture these developments in three interrelated concepts, namely, time–space compression, flexible accumulation and accelerating turnover in production. Harvey observes: 'Improved systems of communication and information flow, coupled with rationalization in techniques of distribution . . . made it possible to circulate commodities through the market system with greater speed. Electronic banking and plastic money were some of the innovations that improved the speed of the inverse flow of money. Financial services and markets (aided by computerized trading) likewise speeded up. So as to make, as the saying has it, "twenty-four hours a very long time" in global stock market' (p. 82). Dyer-Witheford (1999, pp. 130–1) in a perceptive study clarifies this point through Marx's own words that capital 'by its nature drives beyond any spatial barrier' so that 'the creation of the physical conditions of exchange – of means of communication and transport and the annihilation of space by time – becomes an extraordinary necessity for it'.
9. For a recent discussion of the spatial geography of Bangalore's emergence as a 'technopole', see Parthasarathy (2004). Also, for a critique, see Nair (2000).
10. Global media, both print and TV, have been at the forefront of publicising the virtues of these initiatives. Many international magazines have run cover stories or main stories glorifying the major projects launched in India.
11. The project was initially conceived as one which would set up info-shops. Although the project has changed its name more than once, the basic model remains the same: the multi-purpose kiosk attempting to cater for the information needs of the rural population.
12. Interview with K. G. Raj Mohan, Administrative Coordinator, on 26 July 2002.
13. For details of the BIO-Village Project in Pondicherry, see the MSSRF's *Annual Report 2001–2002* (MSSRF, 2002b, pp. 85–8).
14. The literal meaning of this word is 'the ruler of the land'. The government is not always democratically elected, but constitutes the local political and caste elite.
15. Interview with K. G. Raj Mohan, op. cit.
16. Interview with Kankeyan, Village Head, Veerampattinam, on 27 July 2002.
17. Dhanalakshmi, a female volunteer at the *Embalam* centre, quips: 'They ask for Rs 2000 to 3000 as monthly salary. The MSSRF and the Panchayat are not willing to pay. So this centre is managed by women alone.' (Interview with Dhanalakshmi, *Embalam*, 27 July 2002)
18. Satish Jha would call this 'social venture capital'. See also note 1.
19. Interview with Ranjith Khosla, CEO, TARAhaat, Delhi, 6 August 2002.
20. Ibid.
21. Peterson et al. (2001) is a curious study conducted with support from Microsoft and the Markle Foundation, showing unsubstantiated optimism in the success of the project, based primarily on the internal projections of the CSO.
22. Interview with Amol Garg, Field Operation Executive, Bhatinda, 9 August 2002.
23. *Aarthyas* are powerful middlemen in the villages.
24. Interview with Amol Garg, op. cit. The process is indicative of the welter of influences the village elite, particularly the middlemen, can wield in the

implementation of the project. There is a striking similarity here with the process of formation of the MSSRF's kiosks supported by the temple trustees and other village elite.

25. Interview with P. Thiruvengidam, 28 July 2002.
26. Interview with Sakthi, 29 July 2002.
27. Ibid.
28. Nevertheless, such projections abound in the literature on ICT diffusion in rural India. See, for example, Vasagam (1999).
29. For a discussion of the state–industry–civil society interaction in ICT innovations in rural India, see Sreekumar (2003).
30. Interview with Paul Adrien, Director, Department of Agriculture, Pondicherry, 30 July 2002.
31. In 2001 *Asiaweek* reported, on the basis of an interview with M. S. Swaminathan, that 'the Pondicherry Government wants the foundation to extend its network to 208 villages – which to Swaminathan would be a good start. He predicts millions of jobs can be created in rural India as early pioneers spread knowledge to others' (Kapoor, 2001, p. 36). After four years, the project is stuck with 11 to 12 kiosks.
32. QUANGOs are defined as organisations which essentially undertake the responsibility of implementing state-sponsored programmes or public policies, funded by the state but operating at arm's length of the executive without an immediate hierarchical relation with it (Theil, 2002).
33. For a philosophically informed recent critique of Post-Development Studies, see Parfitt (2002).
34. As a rule, groupthink does not admit failures. In practical terms, it promotes the belief that the diffusion of ICTs and enhanced access to the Internet will themselves lead to enormous efficiency gains.

5

The Perils of Excessive Export Orientation

K. J. Joseph

1. Introduction

During the 1980s and 1990s, most developing countries have moved away from import substitution to free trade. The move towards export-oriented growth strategy was prompted by the belief that trade liberalisation resulting from improved resource allocation in line with social marginal costs and benefits, access to better technologies, economies of scale and scope, greater domestic competition and shake-up of hitherto protected industries within a Schumpeterian environment would induce rapid economic growth (Dornbusch, 1992). However, careful observers warn that developing countries should not over-sell trade reforms as a cure for all their economic problems (Rodrik, 1992). While an abysmal trade regime can, perhaps, drive a country into economic ruin, a good trade policy *per se* cannot make a poor country rich (Rodrik, 1992). Nevertheless, countries like India have joined the bandwagon.

The remarkable growth performance of India's information and communications technology (ICT) service sector, especially the software service export during the 1990s under the export-oriented policy regime has been unprecedented, as revealed by a number studies that looked into different aspects of India's software export boom (Schware, 1987 and 1992; Sen, 1995; Heeks, 1996; Kumar, 2001; Arora et al., 2001; Joseph, 2002; Joseph and Harilal, 2001; and Parthasarathi and Joseph, 2002). Looking at India's impressive performance in the software and service sector, D'Costa (2003c) highlights the 'uneven and combined development' taking place in India and calls for a balanced development approach to remedy the contradictions of recent Indian development. This chapter adds to this argument by drawing on the insights from 'Dutch disease economics' to emphasise the opportunity cost of resources being channelled to induce this export boom on other sectors of the economy. An overview of India's software sector follows in section 2. An examination of the linkages of the software export sector with the rest of the industrial sectors is the content of section 3. Analysis of the impact of the software export boom on the product and factor market is undertaken in section 4. The chapter ends in section 5, by calling for a concerted domestic market orientation for the IT sector through a national policy on ICT diffusion.

2. Recent trends in India's IT software and service sector

The value of output of India's software and service sector increased by more than 18 times from less than $0.83 billion in 1994/95 to $15.5 billion in 2003/2004 (NASSCOM, 2004). During the same period, the software for the domestic market increased tenfold. The growth of the sector was fuelled mainly by the export of software and services, which increased by more than 25 times. No wonder that, while India accounts for only about 0.8 per cent of global trade, its share in world IT software service spending is as high as 3.4 per cent. Needless to say, the sustained growth record set by the software and service sector has been unprecedented in India. Such a phenomenal increase, in the light of the discussion in the previous section, may be seen as a combined outcome of various policy initiates and institutional interventions made by the central and state governments and those of the private sector to take advantage of India's comparative advantage in software and services in the growing world market.

2.1 Software services and the Indian economy

Given the high growth of this sector, the share of the software and service sector in total exports increased almost sevenfold (from 3.2 per cent to 21.3 per cent) during 1996 to 2003. Also as a percentage of GDP, the software and service sector today accounts for over 2.6 per cent, as compared to 0.5 per cent in 1996/97 – an increase of over five times (see Figure 5.1). More importantly, India has established credibility in the international software and service firms, which in turn makes the country an attractive location for offshoring of services. The credibility has been established on account of the better Indian software firms' superior performance in quality (see chapter 8). In addition to its contribution to GDP growth, export and employment generation, the IT sector has had some spillover effect, in terms of promoting new entre-

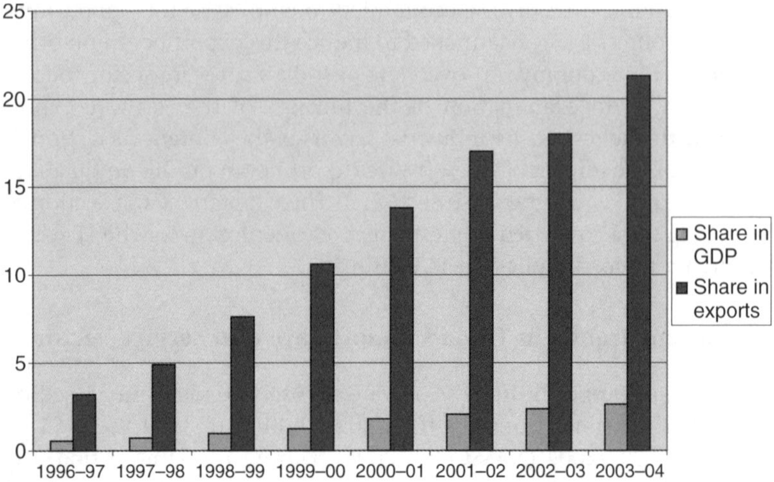

Figure 5.1 Share of software and services in GDP and exports

preneurial and management practices in other sectors of the economy (Arora and Athreye, 2002).

However, even today in 2005, the software and service sector is confined mainly to a few metropolitan cities such as Bangalore, Delhi, Mumbai, Hyderabad and Chennai, thus remaining an enclave with limited linkages to the rest of the economy (Kumar, 2000; D'Costa, 2002a). Various state governments have initiated policy measures to attract software investment in other parts of the country, but the results are yet to be realised.

2.2 Capability building

While some scholars attribute the remarkable performance of India's ICT service sector to benign state neglect (Arora et al., 2001), recent research on the issue in the framework of a national system of innovation has shown that India's performance in the ICT sector is illustrative of the beneficial effects of proactive state intervention, whereby the state policies and institutional interventions made possible the availability of a large pool of skilled manpower on the one hand and institutional infrastructure on the other (Kumar, 2001; Joseph, 2002; Kumar and Joseph, 2004).[1]

The higher export growth during the 1990s notwithstanding, it has been argued that Indian firms operate at the low end of the value chain

(Arora et al., 2001; D'Costa, 2002a). However, several Indian firms have managed to develop and launch a number of proprietary software products. A niche market has been created in banking, financial and accounting software. These include, for instance, I-Flex, which has been used by over 240 financial institutions in 69 countries. Polaris has developed a proprietary retail banking software, Polaris Point, and is joining forces with Bull, France for its marketing in Europe. Banking solutions from Infosys (Financle, Bankaway and Payaway) have been adopted by 22 domestic and 16 overseas banks across 12 countries. TCS has launched packaged software for banking insurance, securities, accounting and health care industries.

2.3 Changing mode of software and service exports

Unlike trade in tangible goods, trade in services can be carried out in different ways. Nayyar (1988) and Sampson and Snape (1985) identify four different modes of service export: (1) trade where the provider is mobile, (2) trade where the receiver is mobile, (3) trade where both the provider and the receiver are immobile, and (4) trade where both the provider and the receiver are mobile. Trade in software and services is carried out mainly through (a) onsite services, (b) offshore services, and (c) offshore products and packages. The underlying distinction between these services is the movement of provider and the receiver.

Onsite services, popularly known in the industry circles as 'body-shopping', occur when the factors move to the site of the receiver. In its extreme form this occurs when software manpower is exported to help solve the user's software-related problems. Here, the firms actually export the manpower and need not necessarily have a complete understanding of the user's requirements. Onsite services might also comprise undertaking the task of software development to solve specific problems. The development and execution of the software are performed onsite. Under this mode of export, it has been argued that the net export earning will be substantially lower than the total export earning because a part of the foreign exchange earned will have to be spent in the importing country itself (Heeks, 1996).

The second mode of ICT software and service export is offshore services. This involves limited movement of both the factor and the receivers. The software is developed offshore according to the specified requirements and exported to the users. This method could be cost effective because the software development process is carried out in the domestic country. At the same time, this calls for more investments in hardware and communication networks.

The third mode of export is offshore packages, or software product development. Here, neither the factor nor the receiver of the service moves. This is also highly capital- and labour-intensive. Additionally, there is also the need for incurring substantial marketing costs. Nonetheless, the net export earning will be higher in this mode of export as compared to others. On the whole, it appears that the extent of net export earning and domestic value addition depends to a large degree on the structure of export. Hence, one of the major objectives of India's policies towards IT software and service exports has been to increase the offshore component.

In the early years of the industry, software and service exports from the country were carried out mostly in the form of onsite development (Heeks, 1996). However, with the setting up of a number of software technology parks, which *inter alia* provided access to modern telecommunication facilities and liberalised policies in the telecom sector (which in turn led to the entry of a number of private sector telecom companies), there has been a significant shift away from onsite development. Today, nearly 60 per cent of exports occur as offshore development (see Figure 5.2). However, the figure also indicates that the share of software products and packages, wherein the profit margins and value addition are known to be higher, has shown a decline in the more recent years. To illustrate, the share of software products and packages drastically declined from about 9.5 per cent in 1995 to 3.2 per cent in 2004. Software products, unlike IT services, are more investment intensive and risky, and there are high entry barriers. Given that Indian firms are smaller than their competitors, a policy would need to facilitate con-

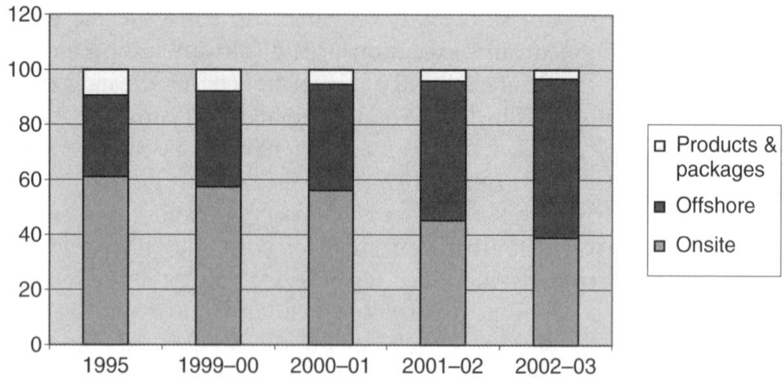

Figure 5.2 The changing mode of software and service exports

solidation, and joint ventures in software product development are crucial for the survival and growth of local industry.

In terms of the destination of exports, the US accounts for 67.8 per cent followed by the UK with 14 per cent. The share of Asian countries is estimated at about 8 per cent, of which almost 3 per cent accounts for the Japanese market (Joseph and Parayil, 2005). Though high regional concentration has to be seen against the fact that today, in 2005, the US alone accounts for over 45 per cent of the total IT spending in the world, the importance of diversifying the export destination and thus reducing the present heavy reliance on a single market cannot be overemphasised.

3. The declining growth of software and the rise of IT-enabled services

It is well known that IT services, in general, are cheaper to deliver in developing countries than in developed countries. This is mainly due to the abundant supply of cheap labour, the major input into the production of services, in developing countries. Since the technology of producing services does not differ significantly across countries, lower wages result in low costs of production of services in developing countries (Bhagwati, 1984). Yet, developing countries, in general, were unable to benefit from this cost advantage because most of the services were embedded in their provider and their export called for the trans-border movement of labour. But the movement of labour, unlike capital, was subjected to a series of restrictions. Through the process of globalisation, which *inter alia* implied the free movement of products and factors, there have been severe restrictions on labour mobility. However, the advances in ICTs have made possible, to a great extent, the 'splintering off' of many of the services from their providers, which in turn led to what is often called the offshoring of services.

No wonder India, with its large pool of skilled manpower, has emerged as a major exporter of IT software and related services like business processes outsourcing (BPO). Our estimates of annual compound growth rates based on data obtained from the National Association of Software and Service Companies (NASSCOM) and the Department of Information Technology show that, while the exports recorded an annual compound growth rate of over 54 per cent during 1996 to 2000, growth rate during the following three years was only of the order of 25 per cent. The period since 2000 also witnessed the emergence of IT-enabled services (ITES) as a major component of software and service exports from the country. In 1999/2000 the total ITES exports was only in the region of $565 million, and it is estimated to have increased to $3.6 billion in 2003/2004. Thus viewed, the recorded growth rate of software exports during 2000/2003 was only of the order of 17.6 per cent. Hence the upward trend in the annual growth rate since 2001 has been mostly on account of the increase in ITES.

Among the different components of IT software and service exports, the only sector with a significant share and recording a growth rate higher than the industry average of 24.5 per cent during 2001/2002 is ITES. Although IT consulting, network infrastructure management and product development and design recorded a higher growth rate, the combined share of all these areas is only about 25 per cent of ITES. As a result, the share of ITES in total software and service exports more than doubled from about 14 per cent in 2000 to more than 29 per cent in 2003/2004, providing employment to about 0.25 million persons.

4. The software sector and its linkages

The observed growth in software services export during the 1990s has created much euphoria and media attention. The main reason for the euphoria appears to be the tremendous growth in export earning and the credibility that India enjoys in the world IT market. If this is the case, one might ask why non-resident Indians (NRIs) who have made significant professional achievements in countries like the US and who also make remittance at least three times that of the software exports do not receive the same credibility in the media? Why is there no excitement about the foreign exchange earned by the NRIs? The software sector brings in foreign exchange thanks to the substantial investments in building up human and physical capital by the state, which, in turn, facilitated the export boom. To this one must add the fiscal concessions provided by the state. But in India there are plantation workers, fish workers, leather workers and diamond cutters and so on who earn substantial foreign exchange without any significant investment by the state towards their skills formation. Yet, why was there no hype about the export earning made by them? Perhaps the underlying reasoning may be best summarised as 'a dollar worth of potato chips is not equal to a dollar worth of micro chips'. Unlike the traditional sectors, the software sector is known for its spillover benefits and linkages with the rest of the economy.

Table 5.1 Indicators of IT diffusion in India's industrial sector (1997)

Industries (2-digit level)	Total no. of factories	% of factories using computers in the office	% of factories with network	% of factories with Internet	% of factories with robots or computers in production
Food products	14,695	13.01	0.84	1.39	0.29
Other food products	8,109	24.17	1.38	2.01	1.64
Beverages and tobacco	8,669	47.81	0.36	0.28	0.14
Cotton textiles	9,227	22.28	0.54	1.87	1.37
Wool, silk, textiles	3,989	49.76	1.25	2.28	0.25
Jute and vegetable fibre textiles	503	16.70	0.40	3.78	0.60
Textiles production including apparel	5,409	51.32	3.18	11.31	2.09
Wood and wood products	3,787	8.98	0.40	0.95	0.24
Paper and paper products	6,304	38.50	1.84	3.73	4.71
Leather products	1,742	37.60	1.89	7.18	0.29
Basic chemicals & products	9,357	50.69	2.91	5.58	2.56
Rubber, plastic & coal	7,597	42.57	2.80	4.01	1.59
Non-met. mineral products	11,376	13.37	0.41	0.95	1.09
Basic metal & alloys	6,915	41.94	0.93	3.69	1.72

Table 5.1 Continued

Industries (2-digit level)	Total no. of factories	% of factories using computers in the office	% of factories with network	% of factories with Internet	% of factories with robots or computers in production
Metal products	8,243	31.68	0.92	2.86	1.01
Machinery & equipment	8,203	44.46	2.12	5.63	2.66
Electric machinery & equipment	5,743	55.77	3.53	10.92	4.89
Transport equipment	3,999	46.96	1.63	7.15	2.58
Scientific equipment	2,243	48.02	4.01	14.00	3.97
Repair of capital goods	2,240	25.89	0.80	1.96	0.36
Electricity	3,644	64.71	0.93	3.10	3.24
Gas and steam	80	75.00	2.50	3.75	5.00
Water works and supply	293	10.58	0.68	1.02	0.68
Non-conventional energy	4	25.00	25.00	25.00	0.00
Storage & warehousing	1,078	0.37	0.37	0.09	0.00
Sanitation	102	0.00	0.00	0.00	0.00
Motion picture etc.	51	7.84	7.84	27.45	0.00
Laundry & others	94	0.00	0.00	0.00	0.00
Repair services	1,966	2.59	2.59	1.12	0.00
All Industries	135,662	34.70	1.50	3.72	1.77

Source: Raipuria, 2002

Analytically, the contribution of ICT could be viewed broadly at two different but interrelated levels. The direct effect refers to the contribution in output, employment and export earnings on account of the production of ICT-related goods and services which are confined to just one segment of the economy (Kraemer and Derick, 2001). The indirect effect may be divided into IT-induced development through enhanced productivity, competitiveness and human welfare (a) on account of the use of this technology by the other sectors of the economy, (b) on account of the emergence of new services enabled by ICTs, and (c) on account of spillover.

Of these, the direct benefits accruing to the economy on account of employment, income and export earning have been well documented (Schware, 1992; Heeks, 1996; Arora et al., 2001: D'Costa, 2003b). But the indirect effect, which involves both positive and negative elements, has not received the attention of researchers that it deserves. However, Kumar (2000) has shown that the IT sector in India has promoted some related service industries like provision of IT maintenance and customisation services, and it has been estimated that these services accounted for about 0.5 million jobs in 1998/99. If the recent evidence is any indication, ITES represent the fastest growing segment of India's IT sector. On spillover benefits, Arora and Athreye (2002) have shown that a potentially important and underappreciated contribution of the software industry is its promotion of entrepreneurship and good corporate governance in India.

But when it comes to the linkages in terms of generating regional growth impulses in a vast economy like India, which is plagued by inter-regional disparities in growth, the sector seems to have fared rather poorly (D'Costa, 2003c). It appears that the growth of the sector has been driven mostly by the export market and it remains an enclave in the economy with hardly any forward and backward linkages. In terms of location, even today in 2005 almost 90 per cent of software development and export activity is confined to five states, and one metropolitan city in each state accounts for the bulk of activity (Kumar and Joseph, 2004). Thus, as observed by Mansell (2002) and D'Costa (2003c), there appear to be few beneficial linkages from the export-oriented software growth strategies pursued so far.

Now let us turn to the issue of the diffusion of ICT to other sectors of the economy, especially the industrial sector. Countries all over the world are engaged in efforts to harness this technology for increasing efficiency, competitiveness and growth of other sectors of the economy. Results of recent cross-country studies indicate that the returns to investments in information technology in terms of productivity and growth are substantial (Pohjola, 2001; Kraemer and Derick, 2001). Pohjola finds the output elasticity of IT capital as high as 0.31 for the full sample of 39 countries and 0.23 in the OECD sub-sample. Another cross-country study by IMF (2001) draws similar conclusions. The country-specific study for Singapore (Wong, 2001) finds that the net return to IT capital (37.9) is about two-and-a-half times higher than that for non-IT capital (14.6 per cent). In this context, it may be pertinent to examine the extent of diffusion of ICT among other industrial sectors in India.

The data presented in Table 5.1 show that only 1.5 per cent of the factories in India's industrial sector are linked to data networks and only 3.7 per cent have access to the Internet. Nonetheless, the available evidence tends to suggest that the use of computers for accounting and management is fairly widespread, with more than 34 per cent of factories reporting use of computers in the office (see Table 5.1). While use of the Internet is low for the industrial sector as a whole, some of the export-oriented industries like textiles and knowledge-intensive industries like scientific instruments are found to be ahead of other industries.

Thus while India can boast of having a vibrant ICT sector, its linkages in terms of harnessing the new technology for improving productivity and competitiveness of other sectors of the economy remain rather low. The limited use of ICT in the domestic sectors of the economy in India has historic roots and must be seen within the context of the country's policies towards the use of computers, especially during the formative years of the computer and electronic industry. On account of the generally held belief that computers are displacing labour in a nation with abundant labour supply, the use of computers, in general, has been discouraged. In 1969 the Ministry of Labour, Employment and Rehabilitation set up a high level committee to advise the government on computerisation. The Parliament resolution setting up the committee noted:

> Government's policy in this regard has been that automation could be introduced on a selective basis. . . . The criteria which determined such selectivity need to be more clearly defined and for this purpose the government have decided to set up a committee (quoted in Subramanian, 1992, p. 13).

The committee favoured a positive approach and recognised the legitimate use of computers in education, science and defence, and even in some commercial and industrial establishments. But the strict regulatory measures that they prescribed were contrary to a positive approach. It was laid down that all the computerisation proposals had to be scrutinised by two experts case by case, and a justification report subsequently furnished.

In 1979, during the first meeting of the reconstituted Electronics Commission, the Cabinet Secretary and the Finance Member reported that computerisation would lead to severe unemployment. As late as in 1984, while the government adopted a liberalised policy towards the computer industry, the approach of the government towards computer usage in the country had not changed much. To a question raised in Parliament in 1984 on the use of computers, it was stated that the policy announced in 1978, which was highly restrictive, continued to be the basis for the induction of computer technology in India.[2] While the then Prime Minister Rajeev Gandhi firmly believed that computers could enhance productivity and national development, the government, at large, was hesitant to openly declare a policy that promoted the use of computers across different sectors of the economy. Though the policy on software export announced in 1986 underlined the need for promoting both export and domestic use of software, it did not highlight any specific measures to promote the use of computers or software in domestic economy.

While the government policy, in general, was lukewarm towards the domestic usage of software, the government nevertheless undertook a series of policy measures and institutional interventions to promote software export. In the face of a growing opportunity in the world market, the private sector, naturally, focused its efforts on the export market without paying any attention to the domestic market.

5. The product market, the factor market and the IT export boom

The macroeconomic implications of the IT export boom may be viewed more precisely from two angles: implications on the product market and on the labour market. The 'Dutch disease' syndrome in trade theory could be used as an analytical tool to reflect on both implications. The central argument of the Dutch disease model is that windfall booms of external income can cause domestic economic problems, even leading to the most unexpected outcome of de-industrialisation of the economy. Literature on the Dutch disease and 'resource curse' underlines the severe backlash caused by a primary commodity boom

(Corden, 1984; Van Wijnbergen, 1984). Dutch disease economics is so named after the experience of The Netherlands in the 1960s, when the country experienced a boom in the discovery of natural gas. The more The Netherlands augmented its natural gas exports, the more in decline its manufacturers of traded goods became. The Dutch disease models were later found to have general applicability in the context of primary commodity exporters, including petroleum, minerals and commercial agricultural goods (Kamas, 1986; Usui, 1996).

The Dutch disease syndrome is explained in terms of two symptomatic effects of an export boom, namely the 'spending effect' and the 'resource movement effect' (Corden and Neary, 1982). The spending of the extra income from the export boom would tend to increase the prices of non-tradeable goods *vis-à-vis* tradeable goods, leading to a real appreciation of local currency and erosion of competitiveness in the tradeable sector. The 'spending effect' refers to the contraction of non-booming tradeable sectors on account of the real appreciation. The tendency for the prices of factors of production and non-tradeables to increase cannot but adversely affect non-booming tradeable sectors that are exposed to external competition. The expansion and increased profitability of the booming sector would draw the mobile factor (labour) out of other sectors and bid up its price. The resulting contraction of non-booming tradeable sectors due to the heightened competition for factors of production is referred to as the 'resource movement effect'. Analytically, the spending effect operates through the product market whereas the resource movement effect is associated with the factor market, more specifically the labour market. Here it may be noted that these two effects are not mutually exclusive nor do they operate in isolation. Instead, they reinforce each other.

5.1 The software export boom and the product market

The implications on the product market operate essentially through price increase and the resultant real appreciation of local currency. Increased export earning leads to a rise in exchange reserves and money supply. The resultant domestic inflation in turn could lead to real appreciation, making the domestic producers of tradeables less competitive. The observed decline in industrial growth and export since 1995 in the context of increased export earning from software services and growing foreign exchange reserves has been cited as evidence of the operation of a software export-induced Dutch disease in India. To empirically gauge the veracity of this argument, we need to place the software export in the context of the external sector in the economy.

Table 5.2 Software exports and India's external balance (US$million)

	1990–91	2001–02
Trade balance	–9438	–12,073
Invisible (net)	–242	14,054
Of which:		
Private transfers	2069	12,125
Miscellaneous services	161	3,774
Software export	128	7,780
Current account balance	–9680	1,351
Total Capital account	7056	9,545
Overall balance	–2492	11,757

Note: All items except software export are given as net
Source: Reserve Bank of India, *RBI Bulletin*, various years and NASSCOM at http://www.nasscom.org, accessed on 15 February 2005

Table 5.2 presents the relative contribution of major items towards major turnaround in India's external sector. The major item in the current account that has contributed towards making current account deficit to surplus is the invisibles. Within invisibles, significant contribution has been made by net private transfers, which recorded an almost sixfold increase during the period under consideration. Another item that has positively contributed to invisibles is miscellaneous services, of which software export is one of the items. It may be noted that in the terminal year the net export earning from miscellaneous services is about 50 per cent of the total export from software, which is only one of the items in miscellaneous services. This takes us to a closer look at the net export earning from software.

Since official data on net export earning from software barely exist, we have compiled the firm level data from the computerised database available at the Centre for Monitoring Indian Economy (CMIE). Table 5.3 presents data on the import intensity of a representative set of firms. It is evident that the total export from the sample of firms increased from Rs 1,243 million in 1992/93 to over Rs 145,303 million in 2000/2001. This represents an annual compound growth rate of over 80 per cent (in Rupee terms) and is marginally higher than the growth rates obtained from NASSCOM data on export for the sector as a whole. The difference may be attributed to the fact that during the earlier years the coverage of CMIE data was relatively less than in recent years. At the same time the imports also recorded a commensurate increase

Table 5.3 Trend in export and import and the import–export ratio for a sample of software firms

Year	Total exports	Total imports	Import–Export ratio
1992–93 (20)	1,243.4	751.8	0.60
1993–94 (34)	2,532.8	1,463.1	0.58
1994–95 (59)	4,297.0	2,383.5	0.55
1995–96 (74)	6,672.7	4,122.5	0.62
1996–97 (87)	8,943.2	5,395.7	0.60
1997–98 (115)	18,404.5	9,607.0	0.52
1998–99 (155)	34,265.2	15,085.3	0.44
1999–00 (238)	84,371.3	41,102.6	0.48
2000–01 (217)	145,303.4	69,027.7	0.48

Note: Total exports and imports are given in Rs million; figures in brackets show the number of firms in the sample
Source: Parthasarathi and Joseph, 2002

(76 per cent). The result has been that import as a percentage of export remained around 60 until 1996/97 and thereafter it declined to reach a level of 48 per cent in the terminal year. If the data presented in Table 5.3 are any indication, one could safely conclude that the net export earning from India's software export is not more than 52 per cent of the gross exports. If the net exports earning were of the order of only 52 per cent, the cumulative contribution of software export to the foreign exchange reserves could be around only 16 per cent of total reserves. It must also be noted that in the estimates presented above, export data have been taken from NASSCOM and may involve substantial overestimation (Parthasarathi and Joseph, 2002).[3] If such possible overestimation is also taken into account, the net contribution of software to the external sector will decrease further. If the evidence so far presented is any indication, one is inclined to infer that the net software export earning from the country is yet to grow into such a level to cause a major adverse effect on the growth and competitiveness of other sectors of the economy through the spending effect. This takes us to an examination of the implications of the software export boom for other sectors of the economy through the resource movement effect.

5.2 The software export boom and the labour market

To begin with, it may be noted that software services is a highly labour-intensive activity. The point may be further illustrated. An examination of the share of labour (wages) in gross value added of 217 software firms in 2000/2001 has shown that it accounted for as much as 66 per cent. Unlike most other labour-intensive industries, the software sector employs mostly skilled labour albeit the level of skill varies according to the nature of activities undertaken by the firm. The available empirical evidence suggests that the ongoing software export boom has to be seen in the context of, among other things, the labour cost advantage that India enjoys (Kumar, 2001; Arora et al., 2001; Arora and Athreye, 2002). The boom in the software sector, needless to say, would have led to an increase in the demand for labour on the one hand and an increase in the wage rate on the other.[4] The increased demand for software professionals, in turn, would have encouraged professionals specialising in areas other than software to migrate to the software sector.

The implications of increased demand for software professionals on the one hand and the rising wage levels and consequent movement of professionals from other areas to software services on the other may be examined in terms of the resource movement effect. To begin with, it may be kept in mind that the resource movement effect is confined to only those sectors of the economy that compete with the software sector for technically skilled manpower. For analytical purposes, the sectors that require the technically skilled manpower are divided into two broad subsectors: (1) those engaged in the production of services, and (2) those engaged in the production of goods like hardware, communication, control instrumentation and so on. The service sector in turn may be divided into two: (1) the software sector, and (2) other service-producing sectors like research, teaching and training.

Following Corden and Neary (1982), the labour market implications on account of resource movement may be illustrated with the help of Figure 5.3, which assumes full employment and other such standards in the specific factors model. Employment in the goods-producing sector is measured from Og and that in the service-producing sector is measured from Os. The labour demand curve for the service sector is drawn as a horizontal sum of that in the software sector and other service sectors. In the figure, the initial equilibrium is shown by the intersection of curve Lg and LS at point e where the wage rate is We. At this point, employment in the goods-producing sector is given by OgL; that in the software sector is CL; and COs shows employment in the service sector other than software.

The boom in the software export sector, on account of exogenous increase in world demand, results in an increase in the demand for labour in that sector (shown by the shift from LS to LS′). This would

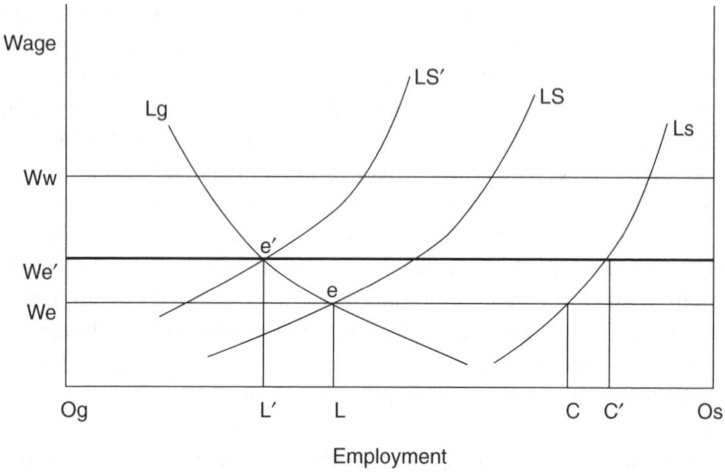

Figure 5.3 The effect of the software export boom on the labour market

cause a rise in the wage rate in that sector (We to We′). The higher wage rate would naturally attract skilled labour, which is mobile, to the software export sector from the other sectors that are competing with the software sector for technical manpower.

Wage rate rises in both sectors. Given that the other sectors do not have a booming world market, firms operating in such sectors will have the following choice: either pay salaries on a par with those of the software firms, or be satisfied with second best by paying lower wages. Given the other pecuniary benefits associated with the software sector, such as employee stock option and opportunity to travel and work abroad and so on, firms in the other sectors are unlikely to attract the best brains even by offering the wages offered by the software sector. Under such conditions, these firms will be forced to employ the second best. Whatever the strategy adopted, there could be an adverse effect on firms operating in sectors which require skilled manpower, leading to a reduction in their employment (from Oc to Oc′), which in turn, by assumption, leads to a reduction in output. On the other hand, given the boom in world demand, employment (and hence output) in the software sector increases (from OsL to OsL′). The observed rise in wage rate is unlikely to have any adverse effect on the software-exporting firms because the world wage rate is still higher than the domestic wage rate (Ww>We′). The wage difference, therefore, continues to give competitive advantage to software-exporting firms.[5]

Our argument that there could be a reduction in the employment and output of other service sectors competing for skilled manpower on account of the resource movement effect could be better appreciated if viewed in the context of the finding by the Task Force on Human Resource Development in IT. It has found that there is a deficit of about 10,000 teachers in software training.[6] How to account for such a shortage? Given that technical manpower for software development and software training is substitutable, and that labour is mobile, there appears to be a preference for employment in the software industry rather than in software training, which is less rewarding. As a result of various initiatives taken in the instance of both the private sector and the government, the annual turnover of engineering degree and diploma holders is estimated at 0.15 million and 0.14 million respectively. Yet the wage premium in the software sector continues and the attrition rate in the software sector still ranges from 25 to 40 per cent (NASSCOM, 2004).

The resource movement effect could also have lead to a reduction in employment in and output of the goods-producing sector, which includes hardware, electronics and communication. The plausible adverse effect of resource movement on hardware production also goes well with the observation made by the National Task Force on Information Technology and Software Development (NTIT&SD, 1998). It observed that 'a steady decline of the IT hardware industry over the 7–8 years due to faulty and deficit policies should be immediately reversed into a growth path through the introduction by a set of policies conducive to growth and international competitiveness' (NTIT&SD, 1998). During 1996 to 2002, the output of the electronics industry recorded a growth rate of only 3.7 per cent, which is found to be only half the growth rate of the manufacturing sector. During the same period, the computer hardware sector recorded a negative growth rate (–1.01) and that of communication equipment 2.2 per cent (Joseph and Parayil, 2005). In general, while the recorded growth rate of the electronics industry during the 1980s was more than double that of the manufacturing sector (Joseph, 1997), with the boom in software exports its performance lagged behind even the manufacturing sector. The Task Force, while attributing the negligible growth of the hardware sector entirely to faulty policies, failed to recognise the labour market linkage between hardware and software sectors.

At this juncture it may be of some interest to explore how the sectors competing for skilled manpower perform in the state of Karnataka, which is known for its pre-eminent position on the software map of

India. Karnataka is the first state in the country to have come up with an IT policy, in 1997. The total number of IT companies in the state, situated mostly in Bangalore, known as the IT capital of India, increased from 13 in 1990/91 to 1154 in 2002/2003, of which 100 or so are foreign firms. Of the 43 companies in India with SEI-CMM Level 5 status, 20 are in Karnataka. Total software output from the state increased from Rs 56 million to Rs 123.5 billion during the above period, accounting for about 29 per cent of total Indian export in 2002/2003.[7] With a view to gauging the implications of the software export boom on other sectors competing for skilled manpower, we have estimated the trend in the share of electrical, electronics and industrial machinery industry in the total investment, employment and output of the manufacturing sector of Karnataka.[8] It is evident that during the 1990s, as software exports were booming, there was a decline in the share of these industries in the total industrial investment, employment and value of output (Figure 5.4). Thus, the boom in software exports has come at a cost in terms of weakening the industrial base of certain skill-intensive and key industries such as the mechanical, electrical and electronics.

From the discussion so far, the following conclusion, though tentative, may be drawn. While the adverse effect on other sectors on account of the spending effect might have been limited, there is some merit to

Figure 5.4 Performance of sectors competing for skilled manpower in Karnataka

the argument that the software export boom had a dampening effect on the cost and competitiveness of other sectors competing for skilled manpower through the resource movement effect. At the same time, we emphasise the need for further empirical inquiry about the implications of the software boom on sectors which compete for skilled labour like IT training, research and development, computer hardware, communication equipment, control instrumentation and so on. Such an inquiry would contribute to informed policy-making in the ICT industry in India.

6. Concluding observations

One of the oft-cited achievements of India during the 1990s was the emergence of the ICT sector, which has shown remarkable vibrancy in terms of output and export growth as well as technological dynamism. These were often cited as an outcome of an export-oriented growth strategy of the government. While these achievements may be considered as direct benefits from the ICT sector, which is no doubt laudable, there arises an issue as to what extent the general public in India benefited from the software boom. It has been argued that the real benefit of the software sector emanates from the linkages and spillover benefits that it generates. It was shown that the positive linkage effect of the software sector appears to be limited, primarily on account of the limited diffusion of IT in the general industrial sectors. From the perspective of regional development, it appears that the software service sector remains an enclave, confined to five locations that account for nearly 90 per cent of software production and export.

Drawing insights from the Dutch disease syndrome, the macroeconomic implications of the software export boom on the product market and factor market have been analysed to assess the effects on resource movement and spending. It was argued that, while the adverse implications on account of the spending effect are likely to be minimal, the software export boom had adversely affected the growth and competitiveness of other sectors competing for skilled manpower. The export-oriented strategy has undoubtedly made India a major exporter and producer of software and ITES. However, the perils of such a strategy are evident from the findings of a study by IMF (2001), which shows that countries using IT tend to reap more benefits than those producing IT. Therefore, mitigating the adverse effect by promoting the diffusion of ICT to different sectors of the economy is essential, and calls for an urgent national policy on ICT diffusion.

Realising the potential of the new technology, the central and several state governments have introduced various projects and programmes to spread as widely as possible the benefits of ICT. Also, various civil society organisations have undertaken a number of initiatives to harness ICT for rural development and poverty alleviation (see chapter 4). The private sector is also involved in some ICT projects aimed at rural development.

While the large number of initiatives have created an impression that India is well on the path to harnessing the new technology for development far ahead of other developing countries, in reality such initiatives have been sporadic, involving the process of learning by doing, and more often than not lacking strategic sense and a national perspective. Many of the experiments remain at prototype level, accomplishing much less than what could have been possible had there been a national system of innovation attuned to effectively harnessing ICTs for development. On the whole, it appears that while India is a major software services provider, its progress in ICT use has been remarkably lacklustre. Indian policy-makers have failed to realise that the social marginal product of a dollar's worth of ICT used at home is bound to be much higher than a dollar's worth of ICT exported.

Notes

1. Changes in the character of state involvement in India's electronics could be traced back to the early 1980s. See Evans (1992) and Joseph (1997).
2. Ram Vilas Paswan, 'Computerisation Policy in the Country', *IPAG Journal*, 11(11), New Delhi, August 1984, p. 742, Lok Sabha Question, No. 4142, answered on 21 March 1984.
3. Based on data obtained from a sample of 155 firms it has been shown that the extent of over estimation in exports during 1998/99 was of the order of 33 per cent.
4. Arora and Athreye (2002) state, 'while the software professional in India earns only a fraction of the salary of his international counterpart, it is still about 20 times the national average'.
5. The prediction that there could be a general rise in wage rate may not hold if the labour market is segmented (which is assumed away here).
6. See 'IT panel sees manpower crisis', *Economic Times*, 16 January 2001.
7. See for details http://www.bangaloreit.com/html/itsckar/itindustriesothercities.htm, accessed 20 February 2005.
8. This refers to the NIC codes 35 and 36 at the two-digit level.

6

Informational Development in Rural Areas: Some Evidence from Andhra Pradesh and Kerala

Jayan Jose Thomas

1. Introduction

This chapter examines the factors associated with the diffusion of information and communication technologies (ICTs) in rural areas, and in doing so it looks at the potential role that ICTs can play in the development of rural areas. Empirical support for the chapter is based on field studies conducted in July/August 2004 in two rural locations in two South Indian states – Kuppam in Andhra Pradesh and Malappuram in Kerala. Various projects and programmes to use ICTs for enhancing developmental opportunities are ongoing in both locations. A major conclusion of this chapter is that ICTs can play a potent role in rural development, but only if the basic obstacles to rural prosperity are removed through radical changes – through land reforms, rural credit and greater state intervention in rural infrastructure and primary education.

Seventy per cent of India's population of more than one billion live in rural areas. Twenty-nine per cent of the total population lives below the poverty line, and 34.6 per cent of Indians are illiterate (see Table 6.1). Indian development today faces a paradoxical situation. On one side we see the phenomenal growth of the ICT industry in some Indian cities, and on the other side the continuing impoverishment of India's rural poor. The case of the state of Andhra Pradesh is very telling in this regard. Andhra Pradesh, and in particular its capital city Hyderabad, has received widespread media attention in recent years for attracting sizeable investments in the IT industry. Over the same period of time, however, Andhra Pradesh has been passing through a phase of unprecedented rural distress, including cases of starvation deaths, distress

Table 6.1 Selected indicators of development: Andhra Pradesh, Kerala and India

Indicator	Andhra Pradesh	Kerala	India
Area (thousand square kilometers)	275	39	3288
Population, 2001 (millions)	76	32	1027
Life expectancy at birth, males, 1993–97 (years)	61.2	70.4	60.4
Literacy rate, age 7+, females, 2001 (%)	51	88	54
Average household expenditure per capita 1999–2000 (Rs/month)	541	810	589
Growth of real per capita state domestic product (% per year) 1990–91 to 1998–99	3.5	5.1	4
Poverty index: head-count ratio (%), 1993–94	26	18	29
Telephones (per 1000 people), rural, 2003	20.3	78.5	14.9
Number of institutions for higher education, 1993–94	850	234	9003
Number of engineering colleges, 1993–94	36	9	323

Source: Drèze and Sen, 2002; Central Statistical Organisation, 1999; Press Information Bureau, Government of India, at http://www.indiastat.com, accessed 5 October 2004

migration of landless agricultural labourers and suicides among farmers and handloom textile workers (Krishnakumar, 2001; Sainath, 2003). The former Chief Minister of Andhra Pradesh, N. Chandrababu Naidu, had won great adulation in the national and international media for the proactive policies of his government concerning the IT industry. However, Naidu's nine-year-old government lost by some margin the elections held in May 2004. The widespread popular discontent against his government's neglect of the livelihood issues of the rural poor is reported to be the major cause for this election defeat (Hindu, 2004).

It is obvious that an enclave-type growth of the IT industry benefiting only sections of the urban population and not making any noticeable change in the lives of vast masses of the rural people is bound to generate contradictions and social tensions (D'Costa, 2003c).[1] In India and other developing countries, it is a challenge for public policy to extend the benefits of ICTs to people in rural areas. The aim of this chapter is to discuss some aspects of this challenge. The next section examines how ICTs help development and other related issues. Section 3 outlines the experiences of Andhra Pradesh and Kerala in the application of information technology to rural development. Section 4 presents the methodology and the rationale for the selection of the study areas for field research. A detailed exploration of the diffusion of ICTs in the two rural areas chosen for this is presented in section 5, and section 6 is devoted to a discussion of the potential role of ICTs in the development of rural areas. The chapter concludes, in section 7 with some recommendations.

2. ICTs for rural development: potential and challenges

There is great optimism in policy-making circles that the new technologies, more specifically ICTs and biotechnology, will benefit the less privileged sections of the world's population. The *Human Development Report 2001* expresses the hope that advances in ICTs and biotechnology 'will lead to healthier lives, greater social freedoms, increased knowledge and more productive livelihoods' (UNDP, 2001, p. 1). According to many of the ongoing policy discussions, the real potential of ICTs lies in the way they can revolutionise the means for transmission of information and communication of knowledge. Information and knowledge are touted as the basis for economic development in the coming decades (World Bank, 1999).[2]

Undoubtedly, information aids the process of development in several ways. Information about markets and technology can help poor households in rural areas. With the opportunities offered by communications technology, a small business unit in a remote town can establish commercial links with distant corners of the world and learn about the latest developments in production technology. In developing countries, poor farmers and craftsmen have traditionally been exploited by middlemen because of their lack of information about prices (Eggleston et al., 2002). Mobile phones and the Internet can be powerful tools for the empowerment of the rural poor by, for example, providing information about prices and markets. *The Economist* (2001) reports that, using mobile phones, some fishermen in Kerala keep themselves informed of fish prices, which fluctuate throughout the day and between the various landing spots, while at sea; in this way, they manage to get higher prices than they used to when selling their catch to the middlemen.

ICTs can be instruments of participation for the excluded sections of society. For instance, members of an isolated community can voice their development needs through the Internet. ICTs have great applications in providing healthcare and education.[3] Applications of ICTs such as multimedia are used in programmes of distance learning for working adults.[4] ICTs can considerably improve the efficiency of governance and state intervention, particularly of local bodies (UNDP, 2001; World

Bank, 1999). However, as Sreekumar has shown (in chapter 4), the problem has to do with realising the potential.

2.1 Challenges and realities of ICTs for development in India

There are several reasons to believe that the above-referred policy discussions are overly optimistic regarding the impact of ICTs on development (see Heeks, 2002; Keniston, 2002; Sreekumar, 2003). According to Heeks (2002), a large part of the present discussion on ICTs and development takes a highly optimistic and technological deterministic view.[5] There are two major concerns with respect to the potential of ICTs for rural development. These are, first, whether people in rural areas have the *capabilities* to use the new technologies, and second, whether people in rural areas have the capabilities to *use* the information provided by the new technologies.

A major factor that impairs the capabilities to use ICTs by people in rural areas is the poor level of physical infrastructure, including deficient telecommunication networks, low penetration of personal computers and poor Internet connectivity. In India, per 1000 people, there are only 50 telephone lines (in March, 2003) and only 0.38 Internet subscribers (in September, 2002). These are averages for India as a whole, which conceal the huge variations between different Indian states and between urban and rural areas.[6] The cost of a telephone or Internet connection in India, in absolute figures and as a share of a household's total income, is considerably higher than the corresponding costs in many other countries including South Korea and China (UNDP, 2001, Table A2.4). Telecom reforms in India from the 1990s have resulted in very impressive gains to the country's telecom infrastructure; yet the reforms so far have not been very successful in extending telecom connectivity to India's rural areas (Chandrasekhar, 2003; Purkayastha, 2002).

The capabilities for producing software and information content that are relevant to people in rural areas and the capabilities required for converting information into useful knowledge are equally important for realising the opportunities promised by ICTs (Mansell and Wehn, 1998; Velden, 2002). For the effective spread of the Internet to rural areas, information content needs to be created in the local languages. It is also pointed out that, while research in technologies is more often carried out in developed countries and caters to the demands in these countries, the specific needs for technologies in poor countries are often neglected (UNDP, 2001).[7] In India, there has been some progress in developing ICTs that are cheaper and more suited to the country's needs, including technologies like simputer, wireless in local loop

(WiLL), voice over Internet protocol (VoIP) and Ka-band satellite communications (Jhunjhunwala, 2001).

The impact of technologies depends on the structure of the society in which the technologies are applied, and this is also true in the case of ICTs (Parayil, 1999). Under the prevalence of a regressive social structure, ICTs can aggravate inequalities – between people in urban and rural areas, between the well-educated and illiterates and between those who possess some form of entitlements and those who do not.[8] Diffusion of new technologies will be very difficult among persons whose educational levels are low. It will also be difficult when social and asset-based inequalities are rampant, as when a person belonging to the scheduled caste (socially oppressed class) is barred from using a village or community Internet server.[9]

High levels of illiteracy and forms of social exclusion based on caste and gender continue to prevail in several regions of India. Literacy rates in India are particularly low in rural areas, among females and among members of socially disadvantaged tribes and castes, especially in states like Bihar, Madhya Pradesh, Orissa, Rajasthan and Uttar Pradesh. Successive governments in India have not been greatly committed to ensuring basic education for the masses: the country has still not achieved the goal of providing free and compulsory education to all until the age of 14 (Drèze and Sen, 2002). Backward social conditions place limitations not only on people's capabilities to use ICTs, but also on people's capabilities to use the information provided by ICTs.

3. ICT and development: the experiences of Andhra Pradesh and Kerala

The field study for this chapter was carried out in July/August 2004 in two rural locations: Kuppam in Andhra Pradesh and Malappuram in Kerala. Experiments to use ICTs for bettering the livelihoods of marginalised sections of the population are ongoing in various places in India. These include the use of personal digital-assistants by health workers in Rajasthan (Reddy and Graves, 2000); imparting functional literacy among working adults in some 80 locations in Andhra Pradesh (Patel, 2002); harnessing the benefits of ICTs to farmers and fishing communities in some villages in Pondicherry (Dugger, 2000); and providing sugarcane prices through a centrally linked IT network for sugarcane farmers in the Warana region in Maharashtra (Vijayaditya, 2000). The Gyandoot e-governance project in Madhya Pradesh and the TARAhaat and Drishtee rural IT projects currently ongoing in Punjab and Haryana are some other such ventures (Kaushik and Singh, 2004). The use of IT for social and economic development experiments in Kuppam and Malappuram has been chosen for the field study.

In April 2002, the Andhra Pradesh government and Hewlett Packard jointly launched a programme to build an 'inclusive community' (i-community) – a community that makes use of ICTs to ensure widespread participation of its members in economic growth in Kuppam.[10] The key feature of the Kuppam i-community project is a network of community information centres, operated by local entrepreneurs who receive financial and technical assistance from the state government, Hewlett Packard and World Corps India (an NGO). This project aims at creating several developmental opportunities in Kuppam.[11]

The state government of Kerala launched the Akshaya project in Malappuram in November 2002. The most important feature of the project is a network of Akshaya information centres operated by local entrepreneurs, with assistance from local self-governing bodies. The project aims to impart IT literacy to at least one member of each household in Malappuram and to meet the information and communication requirements of people in the region. As a result of the project, in August 2003 the village of Chamravattom in the Malappuram district of Kerala secured the distinction of being the first fully 'e-literate' village in India. Basic skills to operate a computer and send e-mail are taught to at least one person in each of the village's 850 households.[12]

Andhra Pradesh has been highly celebrated in recent years for its successes in the IT industry. The state capital, Hyderabad, is home to several large IT firms and public and private sector research centres, and is expected to join the select group of cities all over the world as a top global hub of innovation.[13] Andhra Pradesh sends large numbers of software professionals to the USA every year. Software professionals from Andhra Pradesh accounted for 23 per cent of all Indian software professionals working in the USA in 1998 (Ramachandraiah, 2003). According to the government, the 'IT industry', including IT hardware, IT software and information technology-enabled services (ITES) grew at a compound annual growth rate of over 80 per cent in Andhra Pradesh during the period from 1999 to 2002.[14] Kerala, although capable of more, is, today in 2005, far behind Andhra Pradesh in the growth of its ICTs industry. In 2000/2001, software exports through software technology parks of India (STPIs) was Rs 1.4 billion (or 0.7 per cent of the total software exports through STPIs in India) in Kerala compared with Rs 20.17 billion (or 10.06 per cent of the total in India) in Andhra Pradesh.[15]

However, for all its noted successes in the IT industry, the growth of per capita State Domestic Product in Andhra Pradesh in the 1990s was slower than the national average, and noticeably slower than the corresponding growth in Kerala.[16] The poverty level in Andhra Pradesh is considerably higher than in Kerala (see Table 6.1). In 1999 to 2000, the average household per capita expenditure in Kerala was the highest among all the Indian states, and much higher than the per capita expenditure in Andhra Pradesh (Drèze and Sen, 2002). In 2003, per 1000 population, there were 78.5 telephone lines in rural Kerala compared to only 20.3 in rural Andhra Pradesh (see Table 6.1).

Much more striking are the variations in social achievements between the two states. Kerala has been widely praised for its exceptional achievements in the social sphere, particularly in health and education.[17] However, Andhra Pradesh's social achievements are far from noteworthy. In 2001, the female literacy rate was only 51 per cent in Andhra Pradesh compared to 88 per cent in Kerala. In 1993 to 1997, the average male could expect to live for 61.2 years in Andhra Pradesh compared to 70.4 years in Kerala (see Table 6.1). Kerala has had a relatively successful programme of land reforms and shows relatively low levels of inequality between members of different caste and gender and between people in urban and rural areas; in all these, Andhra Pradesh's track record is very poor. Kerala's development achievements are, largely, the result of public action of democratically elected state governments, of political and social movements and of its people being acutely aware of their rights and responsibilities. Public action of a similar nature is yet to happen in Andhra Pradesh.

4. Methodology and selection of study areas

The field study for this research included discussions with officials of government and non-governmental agencies involved in programmes for the diffusion of ICTs in Andhra Pradesh and Kerala. It also included interviews with entrepreneurs operating the information centres in Kuppam and Malappuram. The core data for this research came from a stratified sample survey of households in selected areas of Kuppam and Malappuram.[18]

In Andhra Pradesh, for administrative purposes, each district is subdivided into 'mandals'. Mandals are further subdivided into 'grama panchayats', and each grama panchayat consists of a number of villages or habitations. The Kuppam area has a population of approximately 320,000, which is spread over five mandals: Kuppam, Gudupalli,

Santipuram, Ramakuppam and Venkatagiri Kota. There are, in total, 99 grama panchayats and 714 villages in the five mandals that constitute the Kuppam area. At present, 13 community information centres (CICs) operate in different parts of Kuppam. Five of these CICs started operations in April 2002, three in September 2003, and another five in April 2004.

After discussions with officials of HP and the mandal revenue office, and entrepreneurs of CICs in the Kuppam area, it was decided to conduct the sample survey in Kadepalle village in Gudupalle mandal and in Venkatepalle village in Santipuram mandal. These villages are located within a three-kilometre radius of a CIC started in April 2002. I prepared a list of all the 141 households in Kadepalle village and 168 households in Venkatepalle village, with the names of each head of household and their ownership of land holdings.[19] This list formed the sampling frame for the survey. A representative sample of 45 households was then chosen from the sampling frame, the representative sample consisting of 20 households from Kadepalle and 25 households from Venkatepalle (representing, respectively, 14.2 per cent and 14.9 per cent of the sample population in the two villages). A stratified random sampling procedure arranged on the basis of the size of land holdings was adopted in the choice of representative samples.

Compared to the Kuppam i-community project, the Akshaya Project in Malappuram is carried out on a larger scale. The Akshaya project covers an entire district (Malappuram), and there are 582 Akshaya centres operating in different parts of the Malappuram district as part of the Akshaya project.[20] In Kerala, for administrative purposes, districts are subdivided into 'taluks', and taluks are further subdivided into villages.[21] There is a total of six taluks and 135 villages in the Malappuram district.[22] After discussions with officials of the Akshaya project and the district and local-level administration and entrepreneurs of the information centers concerned, it was decided to conduct the sample survey in the Karachal ward of Pullippadam village located in Nilambur Taluk; an Akshaya information centre is presently operating in Karachal. I prepared a list of all the 381 households in Karachal with the names of each head of household and their ownership of land holdings.[23] A representative sample of 45 households (representing 11.8 per cent of the sample population) was selected from this sampling frame using a stratified random sampling procedure (with stratification conducted on the basis of size of landholdings).

It may be noted that the Kuppam i-community project and the Akshaya project remain in their initial stages: both projects were begun in 2002. It is still too early to judge the total impact of the projects. Clearly, this was not the aim of the survey. The sample survey focused on how socio-economic conditions affect the diffusion of technologies in rural areas.

4.1 Description of study areas

In the 45 households in which the survey was conducted in Kuppam, there was a total of 215 persons (113 males and 102 females), with 4.8 persons on average per household. The total number of persons in the 45 households in which the survey was conducted in Malappuram was 240 (118 males and 122 females), with 5.3 persons on average per household. The total number of persons above the age of seven in the surveyed households in Kuppam and Malappuram was, respectively, 196 and 223.

In the surveyed households in Kuppam, the primary occupation of the head of household was cultivation in 28.9 per cent of households and agricultural labour in 17.8 per cent of households; in 15.6 per cent of all surveyed households, the head of the household was engaged in cultivation in addition to working as an agricultural labourer. In all, 62.2 per cent of all surveyed households in Kuppam depended primarily on agriculture for their livelihood. A percentage of 13.3 of all surveyed heads of household earned their livelihoods as non-agricultural labourers, which included construction workers as well as workers engaged in other jobs such as making incense sticks. In 8.9 per cent of surveyed households, heads of household ran small businesses, including a small restaurant and a shop for selling vegetables and fruit. A percentage of 13.3 of surveyed heads of household worked in the service sector, doing such jobs as teacher, postman, railway worker and laboratory technician.

Compared with the surveyed households in Kuppam, there was greater diversification of employment opportunities among the surveyed households in Malappuram. In total, 19 of the 45 surveyed households (or 42.2 per cent of all surveyed households) in Malappuram depended on agriculture for their livelihoods. These included five households in which the head of each household was a cultivator, 12 households in which the head of each household was an agricultural labourer, and two households in which the head of each household reported no occupation but the major source of household income was agricultural labour by one or more household members. Heads of 6.7 per cent of surveyed households were non-agricultural labourers, mainly head-load workers; heads of 11.1 per cent of surveyed households were engaged in some small businesses, including working as a middleman for land transactions, running grocery shops and selling books; heads of 8.9 per cent of surveyed households were engaged in other jobs, including estate supervisor, postman and bus driver; and heads of 8.9 per cent of households were reported to be working in the Arabian Gulf countries. In total, in 26.7 per cent of all surveyed households in Malappuram, one or more member (not necessarily the head of the household) worked in the Arabian Gulf countries.

In Kuppam, of the 45 households surveyed, 26 households belonged to a Hindu backward caste (officially, other backward caste or OBC); three households to other Hindu castes; nine households to a scheduled caste (SC), and seven households to the Muslim community. In Malappuram, of the 45 households surveyed, six households belonged to a Hindu backward caste (OBC); one household to another Hindu caste, two households to a scheduled caste or scheduled tribe (SC or ST); 31 households to the Muslim community; and five households to the Christian community.

With respect to the literacy status and level of education, the surveyed households in Malappuram were distinctly ahead of the surveyed households in Kuppam. The literacy rate among persons above the age of seven in the surveyed households in Kuppam was 63.3 per cent; the corresponding figure for Andhra Pradesh was 67.5 per cent. The literacy rate among persons above the age of seven in the surveyed households in Malappuram was 96.4 per cent, higher than the corresponding figure for Kerala of 90.9 per cent in 2001.[24]

5. Diffusion of ICTs in rural areas

To understand the nature of the diffusion of ICTs among the people, four questions were asked of the head of each surveyed household:

1 Have you or any of your family members heard about computers?
2 Have you or any of your family members heard about the information centre in your locality?
3 Are you or any of your family members aware of the services of the information centre in your locality?
4 Have you or any of your family members ever made use of the services of the information centre in your locality?

Table 6.2 presents the responses to these questions by the heads of household in Kuppam and Malappuram.

Table 6.2 Indicators of computer diffusion, Kuppam and Malappuram, 2004

	Kuppam		Malappuram	
Indicators	**number**	**% share of total**	**number**	**% share of total**
Households in which at least one member has heard about computers	28	62.2	45	100.0
Households in which at least one member has heard about the information centre	18	40.0	41	91.1
Households in which at least one member is aware of the services of the information centre	10	22.2	32	71.1
Households in which at least one member has made use of the services of the information	8	17.8	29	64.4
Total number of surveyed households	45	100	45	100.0

Source: Author survey data, July/August 2004

These questions were posed only to the head of each surveyed household. At the same time, opinions were sought and carefully noted from each of the family members present in the household at the time of the survey to verify the responses made by the head of household. It was found that it is more often the younger members who have heard about computers and have used the facilities of the information centre in the locality. In cases where none of the younger members of the family was present in the household at the time of the survey and where the head of the household had insufficient information regarding a family member's familiarity with computers, the responses made by the head of the household may not be correct. Therefore, the results in Table 6.2 may be viewed as responses made by the head of each surveyed household subject to the information available to her or him at the time of the survey.

5.1 Broad trends

Table 6.2 indicates that at least one family member has heard about computers in all the surveyed households in Malappuram and in 62.2 per cent of all the surveyed households in Kuppam. The proportion of the sample of households in which at least one member has heard about the information centre in the locality was 91.1 per cent in Malappuram and 40 per cent in Kuppam; the proportion of the sample in which at least one member is aware of the services of the information centre in the locality was 71.1 per cent in Malappuram and 22.2 per cent in Kuppam.

The use of computers by people in the sample of households was far more widespread in Malappuram than in Kuppam. Questions were put to the heads of surveyed households regarding the use of computers, newspapers, radio and TV by each household member. The results of the responses, given in Table 6.3, show that 30.2 per cent of all persons above age seven in the surveyed households in Malappuram have operated a computer; the corresponding proportion in Kuppam was 7.7 per cent. In Malappuram, at least one family member in 75.6 per cent of all surveyed households has operated a computer. The Akshaya programme in Malappuram has been successful to some degree with respect to its stated aim of imparting e-literacy to at least one person in every household. One or more family members have made use of the services of the information centre in the locality in 64.4 per cent of the sample of households in Malappuram; the corresponding proportion in Kuppam was 17.8 per cent (see Table 6.2). In many surveyed households in Malappuram, more than one household member has taken part in the e-literacy programme conducted by the information centre.

It was learnt that members belonging to the Paniyar community (a socially backward community of scheduled tribes) have taken part in the e-literacy programme in some of the Akshaya centres in

Table 6.3 Penetration of computers and media technologies, Kuppam and Malappuram, 2004

	Kuppam		**Malappuram**	
Type of person	**number**	**% share of total**	**number**	**% share of total**
Persons who read newspapers	76	38.8	181	81.5
Persons who watch TV	148	75.5	159	71.6
Persons who listens to radio	103	52.6	165	74.3
Persons who operate a telephone	94	48.0	187	84.2
Persons who operate computers	15	7.7	67	30.2
Total number of persons who are above age seven	196	100	222	100

Source: Author survey data, July/August 2004

Malappuram. Female members have actively taken part in the e-literacy programme and also made use of other services at the Akshaya information centres in Malappuram. The main instructor in the Akshaya centre in Karachal was a 21-year-old Muslim woman who holds an undergraduate degree in Arabic and has also completed a course in computers.

5.2 Diffusion of computers and socio-economic characteristics of households

The study shows that a relatively high level of literacy and education aids the diffusion of ICTs. As shown in section 4, the literacy rate and educational level in Malappuram are much higher than the corresponding rate and level in Kuppam. In several households in the sample in Malappuram and Kuppam, an important source of awareness about computers was children who attend school, who learn about computers as part of their academic curriculum. There were many surveyed households in Malappuram in which one or more household members have either completed or have been participating in a professional training course in computers, after completion of or along with their formal education.

The association between literacy and awareness about computers is evident in an analysis of the results for the sample of households in Kuppam. As shown in Table 6.4, of the 28 surveyed households that

Table 6.4 Literacy and knowledge of computers, Kuppam, 2004

		computers	information centre
1	Households in which one or more member has heard about	28	18
2	Of which proportion of literates among total household members is more than 50%	19	14
	2 as per cent share of 1	67.9	77.8
3	Households in which no member has heard about computers	17	27
4	Of which proportion of literates among total household members is less than or equal to 50%	15	16
	4 as per cent share of 3	88.2	59.3

Source: Author survey data, July/August 2004 (number of household surveyed = 45)

have heard about computers in Kuppam, 19 (or 67.9 per cent) are 'literate households', or households in which the proportion of literates among total household members is more than 50 per cent. On the other hand, of the 17 surveyed households that have not heard about computers, 15 (or 88.2 per cent) are 'illiterate households', or households in which the proportion of literates among total household members is less than or equal to 50 per cent.

There is evidence for an association between diffusion of computers and level of education of household members. In Kuppam, the average number of persons per household who attend an educational institution is 0.47 among surveyed households in which no household member has heard about computers; the corresponding figure in Kuppam is 1.07 among surveyed households in which one or more household members have heard about computers. In Malappuram, the average number of persons per household who attend an educational institution is 0.73 among surveyed households in which no household member has operated a computer, whereas the corresponding figure is 1.62 among surveyed households in which one or more members have operated a computer (see Table 6.5).

Diffusion of computers among surveyed households is seen to be positively associated with levels of assets. Ownership of land holdings is a

Table 6.5 Penetration of computers, educational level and landholding, Kuppam and Malappuram, 2004

Type of surveyed household	Average number of persons attending an educational institution	Average size of land holdings (in acres)
Kuppam		
Households in which no member has heard about computers	0.47	0.67
Households in which one or more member has heard about computers	1.07	1.93
Malappuram		
Households in which no member operates computers	0.73	0.25
Households in which one or more member operate computers	1.62	0.78

Source: Author survey data, July/August 2004 (number of household surveyed = 45)

good indicator of assets in the surveyed villages in Malappuram and Kuppam. In Kuppam, the average size of landholdings among households in which no member has heard about computers is 0.67 acres. The corresponding figure for households in which one or more members have heard about computers is 1.93 acres. In Malappuram, the average size of landholdings among households in which no member has operated computers is 0.25 acres, while the corresponding figure for households in which one or more members have operated computers is 0.78 acres (see Table 6.5).

5.3 Diffusion of computers and the nature of implementation of rural ICT programmes

It was shown that the proportion of surveyed households which were aware of the local information centre in Kuppam was 40 per cent, much smaller than the corresponding proportion in Malappuram, which was 91.1 per cent (see Table 6.2). Two points may be noted in this regard. First, in Kuppam, the proportion of surveyed households which were aware of the local information centre was noticeably lower than the proportion of surveyed households which were aware of computers (62.2 per cent) (see Table 6.2). Moreover, of the 27 surveyed households which were not aware of the information centre in Kuppam, 11 households (40.7 per cent) were 'literate households', households in which the proportion of literates among total household members is more than 50 per cent (see Table 6.4). Clearly, the low level of awareness about the information centre in Kuppam is not entirely due to the low level of literacy. Some of the differences between Kuppam and Malappuram with respect to the awareness about information centres are related to the nature of implementation of the programme in the two places.

In Kuppam and Malappuram, the information centres are run by small entrepreneurs belonging to the locality. The Akshaya programme in Malappuram is implemented through Panchayat Raj institutions or local self-governments. Therefore, local entrepreneurs running Akshaya information centres work in close association with and receive support from officials of local self-governments, local politicians and social activists. Entrepreneurs of information centres in Kuppam do not receive such local support. Officials of the state government, HP and World Corps India oversee the operation of information centres in Kuppam, but none of these officials has any deep links with the local community. This is one reason why the information centres in Malappuram are more popular with the village population than the centres in Kuppam.

The two-week training programme to impart basic IT skills to at least one member of every household has been highly effective in popularising the Akshaya information centres in Malappuram. In the information centres in Kuppam, according to entrepreneurs running the centres, the main users of computers and the Internet were not local people, but officials of government offices and students of two professional colleges located in Kuppam who mostly do not belong to the village. Interestingly, the i-community project is sufficiently advertised. Large hoardings showing pictures of villagers using computers greet a visitor as she passes through various parts of the Kuppam area. Calendars with pictures of the i-community project were hung on the walls of many of the surveyed households. Apparently, such advertisements have not been very helpful, as members of some of the households in the sample expressed their ignorance about the information centre in their locality even though they were standing in front of the calendar advertisement. This shows that creating awareness about technologies at grass root level is important for the effective diffusion of technologies.

6. Demand for information and the potential role of technologies in rural areas

In both projects described earlier, efforts are ongoing at present to make use of ICTs in a range of areas that affect the daily lives of people: agriculture, health and medicinal care, and education. These projects aim to provide farmers with access to information on new and better agricultural practices and agricultural markets. They also aim to provide information about educational and vocational opportunities and healthcare facilities. In July/August 2004, when the field study was carried out, the Kuppam i-community project was building a 'farming information system', 'tele-agriculture', and an 'expert system' for farmers. It proposes to create an electronic employment exchange and an electronic 'public grievance system'. Through 'Yojanalu', a scheme of the Kuppam project that is already under implementation, people can submit online applications for various government programmes.[25] The Akshaya information centres in Malappuram propose to develop and provide information content on education, jobs, agriculture, health and law. These centres will provide modules on Spoken English, vocational training, personal development, career planning and accounting.[26]

These programmes are yet to be fully implemented in both places. Therefore, it was not possible to ask members of the surveyed households if they had benefited from the information content of any specific programme. Instead, the survey attempted to understand the demand for the information supposed to be provided by the programmes. The questions that were put to the head of each household to understand the nature of the demand for information included the following.

- Are you aware of the prices of agricultural goods produced in your locality?
- What, in your opinion, are the major constraints to agricultural growth in your region?
- Are you or any of your family members searching for jobs?
- Do any of your family members have plans to pursue higher studies?

6.1 Demand for information from rural households

The dependence on agriculture for livelihoods is greater in Kuppam than in Malappuram. One would therefore expect the demand for and potential role of information on agriculture of each surveyed household to be greater in Kuppam than in Malappuram. However, evidence from field studies indicates that this is not the case.

First, 22.2 per cent of all surveyed households in Kuppam did not possess land (see Table 6.6). In 70 per cent of these landless households (or 15.6 per cent of all surveyed households), each head of household was an agricultural labourer. Given the meagre employment opportunities in agriculture, these households survived under distressful living conditions. In Kuppam, households of landless labourers did not have much to gain from information on better agricultural practices or agricultural prices, for the simple reason that these households did not have land to cultivate crops, did not have agricultural crops to be sold in the market and also did not have the bargaining power to convert an increase in agricultural prices into a corresponding increase in wages.

Information on better agricultural practices or agricultural prices was not of great use to a large number of land-holding households also in Kuppam. In Kuppam, 33.3 per cent of all surveyed households owned land that was not irrigated. Agricultural production in many of the land-holding households in the sample suffered from a small size of land-holdings and an absence of irrigation facilities; these households were not selling their agricultural produce in the market. In fact, the survey showed that households selling their agricultural produce in the market as a proportion of all surveyed households was 35.6 per cent in Kuppam.

Table 6.6 Potential role of ICTs, Kuppam and Malappuram, 2004

	Kuppam		Malappuram	
Indicators	**number**	**% share of total**	**number**	**% share of total**
Households which are landless	10	22.2	0	0.00
Households which are aware of the prices of agricultural goods they produce	13	28.9	21	46.7
Households in which one or more members are looking for a job in the organised sector	12	26.7	26	57.8
Households in which one or more members are looking for higher studies	3	6.7	16	35.6
Households in which one or more member is working in a foreign country	0	0.0	12	26.7
Households in which one or more members have visited a government office at least once	37	82.2	42	93.3
Households which have received some loans from formal sector	28	62.2	33	73.33
Total number of households	45	100.0	45	100.0

Source: Author survey data, July/August 2004

According to 51.1 per cent of all households in the sample (or 65.7 per cent of all land-owning households in the sample) in Kuppam, the major constraint to agricultural growth was the absence of irrigation facilities. A substantial number of surveyed households in Kuppam (17.1 per cent of all land-owning households in the sample) noted that non-availability of credit is the major problem they face with respect to agricultural growth. It is noteworthy that there was only a single source of institutional credit in the surveyed villages in Kuppam – a Grameen Bank ('Grameen' means rural) under the management of a religious institution. None of the scheduled commercial banks or co-operative banks operated in the surveyed villages in Kuppam. Absence of adequate physical infrastructure for marketing agricultural goods was another important problem faced by the sample of households in Kuppam. The head of one of the surveyed households, who is the owner of a six-acre agricultural plot and a high school English teacher by profession, pointed out that he was keen to adopt new agricultural practices and venture into the cultivation of new crops, but he was concerned as to how he would be able to sell his agricultural products. Sharp fall in the few years before my field studies in the prices of many of the agricultural commodities produced in the village, particularly tomatoes, is another issue of great concern to the cultivating households in Kuppam. Given the persistence of fundamental constraints to agricultural growth as discussed above, there was only limited demand from the surveyed households in Kuppam for information on better agricultural practices or agricultural prices.

The demand for information on agriculture from the sample of households in Malappuram was high compared to the sample of households in Kuppam. In Malappuram, 46.7 per cent of all surveyed households were aware of the prices of agricultural goods produced in their village; the corresponding proportion in Kuppam was 28.9 per cent (see Table 6.6). Most of the surveyed households in Malappuram regularly keep track of the news about agriculture from newspapers, TV and radio. The fluctuations in the price of rubber, the major agricultural produce in the village in Malappuram, was a topic of everyday conversation in public places in the village. Many households in the sample in Malappuram were keen to know about better agricultural practices and venture into the production of new crops such as vanilla.

What are the reasons for the greater demand for information on agriculture from the sample of households in Malappuram than from the sample of households in Kuppam? First, unlike in the case of Kuppam, all households in the sample in Malappuram possessed some plot of land, which was more than five cents (one-twentieth of an acre) in the case of 97.8 per cent of the sample and was more than ten cents in the case of 84.4 per cent of the sample. Most households in the surveyed village in Malappuram (as in the rest of Kerala) had a well dug in their plot of land; and, given the plentiful rainfall in the region, availability of water was not a major constraint to agricultural production in Malappuram. Even a household in possession of only five cents of homestead land could cultivate vegetables or some other agricultural crops. The major agricultural crops in the village in Malappuram – rubber, coconut and areca nut – are high value-adding commercial crops. The surveyed households in Malappuram received credit from more than six different sources of institutional credit, including scheduled commercial banks and co-operative banks. In respect of the physical infrastructure for marketing agricultural products, the sample of households in Malappuram was found to be far better placed than the households in Kuppam. Agricultural workers in Malappuram, like agricultural workers in the rest of Kerala, have sufficient organisational strength to convert an increase in prices of agricultural crops to an increase in wages. Therefore, in Malappuram, even agricultural labourers who do not produce any agricultural crop to be sold in the market are still interested in knowing about agricultural prices.

The demand for information about employment opportunities and higher studies was significantly higher among the sample of households in Malappuram than among the households in Kuppam. One or more household members were searching for jobs in 57.8 per cent of the surveyed households in Malappuram; the corresponding proportion in Kuppam was 26.7 per cent. Members in 35.6 per cent of the surveyed households were interested in pursuing higher studies in Malappuram compared to only 6.7 per cent in Kuppam (see Table 6.6).

One or more members in 26.7 per cent of all sample households in Malappuram were working in Arabian Gulf countries (see Table 6.6). In comparison, a family member was working outside the village (but not outside the country) in only 4.4 per cent of the surveyed households in Kuppam. In Malappuram, 84.2 per cent of all persons above the age of seven in the surveyed households have used a telephone, compared to 48 per cent in Kuppam (see Table 6.3). To the households in Malappuram, the Internet offers a cheaper and much-needed means of communication with their dear ones in the Arabian Gulf region.

6.2 Demand for programmes in e-governance

There is evidence for greater interaction between the government and the people among the surveyed households in Malappuram than among the surveyed households in Kuppam; therefore, measures for e-governance will produce greater benefits in Malappuram than in Kuppam. The proportion of surveyed households in which one or more household member has ever visited a government office is 93.3 per cent in Malappuram and 82.2 per cent in Kuppam. The proportion of surveyed households which have received loans from formal financial institutions is 73.3 per cent in Malappuram and 62.2 per cent in Kuppam (see Table 6.6).

Most of the proposed initiatives in e-governance are aimed at reducing the time and transaction costs involved in government–people interaction. Two points may be noted in this regard. Among the surveyed households in Kuppam and Malappuram, 62.3 per cent and 68.9 per cent respectively answered that they did not face any difficulty – and made two or less trips to a government office per transaction – while interacting with government offices. An explanation for this relatively high degree of satisfaction in transactions with the government offices is that the opportunity cost of the time spent in transactions is likely to be low for people in rural areas. Additionally, members of the surveyed households in Kuppam and Malappuram pointed out that their interaction with the government happens mostly at times when the government introduces new schemes or programmes such as food for work. This suggests that the programmes for computerisation of government–people interaction will have less demand in rural areas than in urban areas and, of course, the demand for these programmes is limited by the schemes and programmes that the government introduces.

The observed differences between Malappuram and Kuppam in the demand for and potential use of information are, as evident from earlier paragraphs, related to the differences in socio-economic structures between the two locations – in turn, due to the differences between Kerala and Andhra Pradesh in bringing about positive transformation in the countryside. The association between existing socio-economic status and demand for information is clear also when we extend our analysis to the various land-owning categories of sampled households in Malappuram and Kuppam. The proportion of households which were aware of agricultural prices, whose members were searching for jobs and whose members were interested in pursuing higher studies is most in the highest land-holding category in both Kuppam and Malappuram.

Landlessness among labouring households is a continuing feature in several parts of rural India. A relatively successful programme of land reforms was implemented only in two Indian states, Kerala and West Bengal. There was a significant decline in public investment in agriculture and rural infrastructure in India, particularly during the period of economic reforms since 1991 (Ramachandran and Swaminathan, 2003). There was also a decline in the volume of rural credit disbursed by banking institutions over this period of time (Ramachandran and Swaminathan, 2002). Landlessness, absence of irrigational facilities and institutional credit, and illiteracy are among the fundamental constraints to the freedom of labouring women and men and to the growth of agricultural incomes in the Indian countryside. The evidence presented earlier showed that these fundamental constraints also significantly affect the demand for information and the usefulness of ICTs in rural areas. ICTs can play an important role in rural development, but only if the basic obstacles to rural prosperity are removed through radical changes – through land reforms, revitalisation of rural credit and greater state intervention in rural infrastructure and primary education.

7. Conclusion

There is abundant optimism today regarding the potential benefits of ICTs in the development of rural areas. This optimism originates from the belief that ICTs are a revolutionary means for the transmission of information, and information is vital to the very process of development. Information about prices, markets and opportunities will enhance incomes and empower the less privileged and the rural population, so the argument goes. This chapter examined the validity of this argument by examining two ICTs diffusion projects, in Andhra Pradesh and Kerala.

The study shows that people's capabilities to use ICTs are associated with the existing level of socio-economic development. The diffusion of ICTs is faster among literates than among illiterates, and also faster the greater the level of education and the greater the level of assets. Compared to Kuppam, the diffusion of ICTs was more widespread in Malappuram, where ICTs are used by women in large numbers and also by the socially backward classes. The Akshaya programme in Malappuram elicited greater public participation as it was implemented through local self-governments with support from social and political activists in the locality.

It has been found that both the capability to use information and the demand for information are crucially dependent on the level of social and economic development. There is no doubt that information about better agricultural practices helps the expansion of rural incomes. Such information, however, will be irrelevant to landless agricultural labour households or to land-holding households who lack irrigation facilities and institutional credit. There will be little demand for information on jobs and higher studies from the less educated.

The evidence presented in this chapter has great implications for the developmental discourse in India and other developing countries. It shows that while ICTs offer great possibilities, they represent no panacea for development. In fact, ICTs can offer little help as long as fundamental constraints to development – including unequal distribution of assets and illiteracy – persist. The current policy of enthusiasm regarding information technology should in no way detract attention from the much-needed policy interventions in income growth and distribution, including land reforms and public investment in health and basic education.

Another policy implication is that the focus on programmes to introduce ICTs in rural areas should not only be on supply-related factors like physical infrastructure for ICTs, but also on creating demand for ICTs and for the information provided by ICTs.

Notes

1. According to D'Costa (2003b), an enclave-type growth of the IT industry has led to 'uneven and combined development' in India.
2. World Bank (1999) notes that knowledge is the basis for economic development and that it is because of information and knowledge gaps that poor countries lag behind richer ones.
3. On the use of ICTs in the delivery of health services, see Chandrasekhar and Ghosh (2001) and Reddy and Graves (2000).
4. See, for example, Patel (2002).
5. A less optimistic and socially deterministic view would be more realistic, according to Heeks (2002).
6. See Thomas (2005a), Table 8.5.
7. Mowery and Oxley (1995) argue that developing countries must build national systems of innovation to facilitate transfer of technologies from advanced countries and to create technologies that are appropriate to the needs of developing countries.
8. For a review of some of the arguments of this nature, see Schech (2002).
9. UNDP (1999) reports that people who access the Internet are more often the better educated and the rich, men rather than women, and younger rather than older people. A study by Arun and Arun (2002) shows that ICTs may reproduce or even intensify many of the broader gender inequalities.
10. See www.hp.com/e-inclusion/en/project/kuppam.pdf, accessed 9 June 2004. See also Dunn and Yamashita (2003).
11. The assembly constituency of Kuppam has been returning Chandrababu Naidu to the State Legislative Assembly continuously from 1989 – and even in the most recent elections held in 2004.
12. See the editorial in *The Hindu*, 9 August 2003.
13. *Wired* magazine, cited in UNDP (2001), p. 38.
14. 'Information and Communication Technology Policy of the Government of Andhra Pradesh 2002', Government Order Ms. No. 27, dated 27 June 2002, available at www.ap-it.com/ictpolicy02.pdf, accessed 9 June 2004.
15. See http://planningcommission.nic.in/plans/stateplan/sdr_punjab/sdrpun_ch13.pdf, accessed 5 October 2004.
16. See Thomas (2005b) for details on Kerala's economic and industrial growth over the decades.
17. For details on Kerala's development experience, see Ramachandran (1996) and Parayil (2000).

18. The methodology for the field study was inspired by Ramakumar (2004), Pais (2003) and Rawal (2001). It also benefited from an ongoing study conducted by V. Surjit of the Indian Statistical Institute, Kolkata.
19. As part of the Integrated Child Development Services (ICDS) Project, which is being carried out in the Kuppam area, 'Anganwadi' (nursery) teachers conduct a door-to-door survey of all households in their village, collecting socio-economic information about households as well as health information of children. I have used this information collected by Anganwadi teachers in Kadepalle and Venkatepalle in the preparation of the sampling frame.
20. See http://www.akshaya.net/itmission/akshaya/vitem.asp, accessed 5 October 2004.
21. In Kerala, a village is an administrative sub-division of a Taluk. For purposes of local self-governance, Malappuram district (district panchayat) is subdivided into 14 block panchayats and 100 grama panchayats.
22. The six taluks are Nilambur, Ernad, Perinthalmanna, Tirur, Ponnani and Tirurangadi.
23. As part of the Akshaya project, entrepreneurs running Akshaya centres have prepared a list of all households with their relevant socio-economic information. I have used this information in preparing the sampling frame.
24. The Census of India 2001 at www.censusindia.net, accessed 5 October 2004.
25. Discussions with officials of the Kuppam i-community project; see also http://www.hp.com/e-inclusion/en/project/kuppam.pdf, accessed 4 April 2004.
26. Department of Information Technology, Government of Kerala, 2004.

7

Building a 'Soft Region' on Hard Legacies: The Informational Economy in Andhra Pradesh, India

Kyle Eischen

1. Introduction

Since 1990, the state of Andhra Pradesh has emerged as an important side of the South Indian 'Silicon Triangle', the other two sides being Karnataka and Tamil Nadu. A combination of global labour networks, foreign capital, local entrepreneurs and regional government initiatives has propelled the region from virtual non-existence to an increasingly central node of software and information technology services in the global economy. These patterns – building upon past national legacies and an expanding global industry – are indicative of both the broader changes within India over the last decade and the potential for 'informational' industries to thrive in unexpected and new environments. Yet, in spite of success within the global industry, the 2004 parliamentary elections witnessed the electorate's resoundingly removing from elected office the chief architects of the regional policy in Andhra Pradesh – the Telugu Desam Party, led by Chief Minister Chandrababu Naidu. Equally important, the electoral change extended across India to both national and state government leaders that had presided over and emphasised India's emergence as a key actor within the global information economy.

This contradiction – success in building an informational industry and failure with the electorate – suggests it is important to re-examine the nature and impact of informational development in India, in general, and in Andhra Pradesh, in particular. This chapter addresses this issue through two interconnected aspects surrounding growth during the 1990s. First, while the IT industry has grown, it still remains unclear in most analyses exactly what spurred such growth and in what context such success can be defined. Simply success for specific firms or even the regional industry does not necessarily signal widespread and general economic development. There were significant market, policy and industry shifts during the 1990s that need to be correlated with ground level transformations. Shifts in policy and industry formation, in particular the emergence of IT-enabled services (ITES) over the last decade, signal the limitations of earlier models and the need to move towards new solutions by both policy-makers and firms. That such shifts still did not significantly alter the outcome of the recent elections presents an opportunity to understand the possibilities, in particular, and the political economy ramifications, in general, of information capitalism.

The relative success of the Andhra Pradesh IT industry development initiative – and the emergence of Andhra Pradesh and its capital city of Hyderabad as players in the global economy – challenges several widely held conceptions of the state's role in promoting meaningful development for the vast majority of the people under a globalised and neoliberal information-driven economy. Essentially, Andhra Pradesh attempted to move directly from a rural to an information economy, but it is doing so by skipping any form of industrialisation, including high-technology manufacturing. Understanding the resources that the region has drawn on to both formulate and establish this development initiative is central to knowing how and whether such policies can be replicated extensively throughout India or other regions in the world and connected to widespread economic and social development that is fundamentally different to previous national variants.

This chapter begins by outlining the overall environment in which the Andhra Pradesh IT industry emerged. This forms a basis for defining the nature of the success of the regional industry as well as its overall impact on the regional economy. The subsequent sections consider the shift from software to ITES within the regional economy and the rationale behind the shift. This leads to revisiting the recent elections and the question of informational development's ability to generate widespread growth. Finally, general lessons for India and other emerging informational regions are considered.

2. Building an 'information economy' within global and national spaces

The emergence of Andhra Pradesh from virtual anonymity to a central location of software investment and production within the global economy is the result of both plan and circumstance. For most of the 1980s or even early 1990s, few outside observers would have foreseen the possibilities of development in Andhra Pradesh in general, let alone development centred on one of the most dynamic and leading industries of the coming century. The regional government under Naidu deserves recognition for understanding, much better and earlier than most other analysts, the opportunity and requirements for building an informational industry like software. The industry emerged from an opportunity that stemmed from a unique combination of economic crisis, development legacies and unique production features within the global software industry.

The early 1990s presented Andhra Pradesh with a double economic crisis. Nationally, the balance of payments crisis created new demands for economic liberalisation, concluding a slow opening of the Indian economy that had begun in the early 1980s. In broad terms, the liberalisation simultaneously opened India to the global economy through new financial and export policies and decentred the national economy by opening spaces for state governments to promote growth more directly. On the regional level, Andhra Pradesh faced an economic crisis of its own making, with rural development, agriculture and electricity subsidies and policies, in particular, pushing the state to the edge of bankruptcy. Combined with this, economic policy as usual was not an option. The national economic crisis limited Andhra Pradesh's ability to turn to the federal government for support, and the rural policies that had been used to promote growth were no longer financially feasible.

Equally important, however, the legacy of both state and national development models and policies left the region singularly unprepared for competition in a liberalised, global economy. The state was over 60 per cent rural, a large part of the state income was agriculturally derived, literacy rates hovered around 50 per cent and there was no significant manufacturing industry that could serve an export market. In comparison with fast-growing states like Tamil Nadu and Maharashtra, Andhra Pradesh was 25 per cent poorer on a per capita basis, 30 per cent less literate and had 46 per cent lower investment rates (World Bank, 2004). On both the social and economic indicators, Andhra Pradesh's performance was rather dismal (see the previous chapter by Thomas for Andhra Pradesh's social and economic indicators).

Yet, a central dichotomy of Andhra Pradesh is that its overall dismal social and economic backwardness was not an overwhelmingly limiting factor for the promotion of the information industry. In actuality, the skewed nature of national development has provided many of the most positive resources around which the regional strategy was implemented. One Andhra Pradesh-specific aspect of this combination is the case of skilled labour. While lacking broad basic education levels, Andhra Pradesh does produce a significant quantity of high-quality skilled labour, especially in the IT field, a large proportion of which migrates to the United States. Combined with limited local opportunities, these professionals were an important resource upon which the regional information strategy was built. The central point to keep in mind is that the legacies of national development were only failures when judged from a traditional state-centred industrial viewpoint. Under national liberalisation and the expansion of software production globally in the early 1990s, these distortions became strengths around which informational-focused development could be considered as not only desirable but also possible given the demands of the global market.

3. Building new institutions for development in Andhra Pradesh

Andhra Pradesh in many ways established the model for informational development in India. As a late mover in the software industry and beginning from a comparatively disadvantageous starting-point, Andhra Pradesh highlights three facets of the regional initiative that are now extensively emulated within India: the replication of national resources on the regional level; the building of unique regional capacities; and the fostering and capturing of private initiatives. The key aspect of each of these initiatives is that together they form an organic whole that was essential to developing the viability of the information-focused strategy in Andhra Pradesh. Overall, the Andhra Pradesh regional strategy focuses on the development of a 'hi-tech habitat' that develops the infrastructure, partnerships, regulation, financing and social norms that will create an innovative region over the course of the next two decades (Government of Andhra Pradesh, 2000).

3.1 Replicating and capturing national institutions and resources

Much of the effort in Andhra Pradesh during the 1990s focused on replicating key national institutions or capturing existing national capacities to fulfil regional goals. Much effort was focused on developing regional models of national institutions and capacities that have spurred development in other regions in India. While national legacies did provide key resources to initiate development, there were gaps in this inheritance when compared with other regions. In addition, existing national institutions became captured and transformed into *de facto* if not *de jure* regional institutions. Key developments along these lines have been:

- *Software Technology Parks, Hyderabad*
  Originally begun as a national programme for targeted investment in telecommunications across India to support existing software firms, the STP in Andhra Pradesh has become an integral part of the regional initiative, promoting software investment in Andhra Pradesh in competition with other regional STP centres in India. It has also become a leading partner with the state in expanding STP services (telecommunications, single window clearance) to new urban regions throughout Andhra Pradesh, hoping to capture investments and small local software start-ups funded through new state-sponsored venture capital.
- *HI-TEC (Hyderabad Information Technology and Engineering Consultancy) City*
  The HI-TEC City is the premier regional investment, the anchor around which a regional information technology cluster ('Cyberabad' in local marketing language) can develop. The $850 million software park is explicitly modelled on the Electronics City in Bangalore that developed around the original STP. It has been central to attracting new investments, including Microsoft, Metamore and Oracle, since its opening in 1998. Additionally, companies such as Infosys, Wipro, Infotech and Vanenburg have undertaken new or expanded investments near the HI-TEC City.

  Structured as a land for equity swap (the regional government provided 158 acres of land in return for an 11 per cent equity stake), the venture is a public–private partnership between the Andhra Pradesh Investment Corporation (APIC) and L&T Engineering. The overall planning began in early 1995 just after the assumption of Naidu as the Chief Minister. HI-TEC City forms a hub in a green-field site west of the city, where urban land policy has been used to geographically link new investments in software campuses, universities, research centres, new housing initiatives and telecommunication and transportation infrastructure around which the software industry will expand region-wide.
- *Indian Institute of Information Technology (IIIT)*
  The IIIT is a paired investment to the HI-TEC City. The goal is to build from the ground up an institution equivalent to the six exist-

ing Indian Institutes of Technology and Science found in other regions throughout India.

While initially the premier institute was a national initiative, it has been formally organised as a non-profit education society autonomous from national and state governments in order to bypass restrictions on curriculum and admissions. Renamed the International Institute for Information Technology (being unable to use the original IIIT designation of the government), the IIIT has actively sought private support and endowments, drawing investments from many of the same firms investing in the HI-TEC City. As an example of this, both Motorola and Microsoft have endowed professorships, and Satyam has funded a language technologies research lab.

Curriculum and training are supplemented by private schools and training institutes, including the Microsoft School of Software Technology, the Oracle School of Advanced Software Technology, Motorola School of Communication Technology, the PSINet School of Excellence in Software Development Methodologies, the CISCO Academy Training Centre and the IBM School of Enterprise Wide Computing. While the private academies reside on the campus, they are autonomously run and developed, though some of the teaching staff assume adjunct positions at the IIIT.

- *Indian School of Business (ISB)*
  The Indian School of Business is located next door to the IIIT as a local variant of the five Indian Institutes of Management. It aims to 'groom the next generation of leaders for India' and promote the general transformation of business in India through innovation and entrepreneurship. The school is the idea of Rajat Gupta, a managing director of McKinsey & Co. The Rs 3600 million ($80 million) school came to Andhra Pradesh when Maharashtra insisted that it must reserve seats for scheduled caste students. The school is formed in partnership with the Wharton School of Business, the Kellogg Graduate School of Management and the London Business School. It is in conception and reality a reflection of non-resident Indian networks.
- *The Institute of DNA Finger Printing*
  This is a new, nationally funded research institute specialising in basic research and development in DNA fingerprinting, molecular diagnostics and bioinformatics. It is also the bioinformatics national node for the European Molecular Biology network (EMBnet), the national centre for bioinformatics services, including bio-molecular sequence databanks, macromolecular structure databank and

genome databases, as well as a graduate school in the life sciences. The institute as such fits in perfectly with the regional focus on developing knowledge-based industries backed by institutions to generate new research and skilled human resources.

- *Civil Service and Management Skill to Run Government Programmes*
  One of the more notable aspects of the development initiative in Andhra Pradesh has been the implementation of innovative government projects developed and implemented by career bureaucrats and locally based national research institutes. The capturing and directing to local needs of these national management and research schools is one of the key factors in establishing the management skills to direct the regional development initiative.

3.2 Building regional institutions and capacity: understanding knowledge-based production and innovation

The adaptation of existing resources and the building of local models to fill in gaps in the national legacies have been combined with unique regional initiatives. One of the central features of these projects is their search for global best practice. Regional planners have freely borrowed or adapted programmes from successful initiatives in India, South-East Asia, Europe and the US. They have also created programmes that are themselves models for other regional initiatives. Key programmes include:

- *AP Biotechnology Park*
  Like the HI-TEC City, the park is designed to be a focus for commercialisation of biotechnology research and manufacturing as well as providing a basis for the application of biotechnology to agricultural needs within the state. The park is set up to generate synergy with other technology-based companies and public research organisations, creating opportunities for networking, collaborations and technology exchange.
- *Venture Capital*
  The government created local VC funding totalling Rs 150 million ($3.4 million)[1] that will be privately managed. This has been supplemented by regulations that will allow banks to invest in VC funds and tax-pass through benefit to avoid double taxation, irrespective of the form of venture capital.
- *Regional 'SMART' Initiatives*
  Not only has the regional initiative captured needed management and research resources from national institutions, it has focused them

on programmes that have been integral to creating an overall regional strategy in Andhra Pradesh. This combination of national skills, local domain knowledge and international best practice used to apply IT to very real problems constitutes one factor that separates Andhra Pradesh from other regional efforts. The regional government philosophy is framed around the concept that 'IT is SMART', meaning that IT is a political necessity to make government Simple, Moral, Accountable, Responsive and Transparent. Here the conception of IT is much broader than just an industry, viewed as an integral part of governance and society overall. Key programmes include:

- *Chief Minister's executive information system.* An online database implemented in 1998 to provide information on electricity, water, health, finance and so on, which decision-makers use to gauge the daily progress of various infrastructure projects and government services.
- *AP state-wide area network.* A network backbone implemented in November 1999, APSWAN provides a 2-Mbps fibre optic link between the state government and its 25 districts.
- *Computer-aided Administration of Registration Department (CARD) programme.* A state-wide system piloted in 1998 to streamline the process of granting land titles and paying land taxes. The process, inherited from the British and still done by hand all over India, had previously taken months to complete, but is now finalised while citizens wait. The final implementation at all 387 state-wide offices was completed in March 2003.
- *Fully automated system for transport (FAST) programme.* Electronic system piloted in May 2000 for issuing drivers' licences at four centres in Hyderabad and 32 state-wide offices. The scale implementation to 35 sites in Hyderabad was completed in January 2002.
- *eSeva programme.* Piloted in December 1999, the programme established centres for utility bill and tax payments and for applications to receive government licences and services. The scale implementation of 27 sites in Hyderabad was completed in August 2001.
- *Smartgov programme.* Launched in November 2002 to create a centralised database system integrating the information of each government division to improve efficiency and transparency. Currently with 30 centres in Hyderabad, the programme aims to expand to 48 new towns.

3.3 Private initiatives and networks: capturing synergy

In addition to these government initiatives, there have been private initiatives that have supplemented, and in some cases provided models for, regional initiatives.

- *Satyam Corporation*
  As one of the leading software firms in India, Satyam has provided a general pattern of development that the government has explicitly followed. The company has built up an entire infrastructure, in essence a corporate city, devoted to software development that is a model of the software and services business within the region and the country. This is combined with a business culture that is both distinctly local and global simultaneously. Satyam was the second Indian firm listed on the NASDAQ but retains its global headquarters in the countryside, an hour-and-a-half outside the capital city of Hyderabad. It seeks to be a global company, yet has been an active and even leading force in the development of a regional software industry.
- *ICICI Knowledge Park*
  This is a private R&D park and network, developed by the International Credit and Investment Corporation of India and designed to facilitate research and technology exchange between universities, research institutes and the private sector. It has linkages with all the major national institutes in Andhra Pradesh, all of the IITs and the IISc, the International Centre for Genetic Engineering & Biotechnology, Delhi and the National Institute of Immunology, Delhi.

4. Evaluating the success of the informational strategy

Given these extensive efforts by the state government, the open question is how successful these policies have been. There are two aspects of success that are important to focus upon. First, there is the growth of the industry and its sustainability. Simply, the question is, how large and embedded in the regional economy is the information sector? The second question is, what have been the impacts of these efforts in addressing the economic and developmental challenges facing the state in the 1990s?

The answer to the first question has already been suggested. As the regional statistics indicate, the software industry's growth was extraordinary during the 1990s: from 15 companies to over 1100 registered with the STPH, from less than a million to over a billion dollars in exports, and consistently within the top four states in the country with 8 per cent of all Indian software exports (see Table 7.1).

These export and industry numbers are matched by over 75,000 jobs directly linked to the industry (see Table 7.2). The vast majority of jobs were created after 1999, with employment rising from 15,000 to its level at the time of writing in less than five years. Equally important, several indirect positions were generated for each of these jobs.

In terms of the robustness and sustainability of the industry, the most significant indication is the level of growth after the end of the 1990s boom. The boom itself was not only linked to the Internet boom in the US, but was also deeply affected by the Euro conversion shift and the Y2K efforts. That the regional economy has continued to grow even after these combined market forces ended represents a serious and sustained competitive position for the regional industry.

Beyond this growth, which is substantial, Andhra Pradesh helps elaborate on how sustainable regional initiatives require understanding software (or information-based industries, more broadly) not as a product or an industry, but as a production process that epitomises knowledge valorisation within a global, networked innovation-driven environment. Overall, the industry's sustainability involves: (1) creating a regional innovation system that supports the constant evolution of products, services and firms (Malecki and Oinas, 1999), and (2) understanding the process by which such a regional innovation system connects local and global knowledge in circuits that apply to multiple facets of regional development.

As such, the Andhra Pradesh strategy has focused not only on attracting foreign investment, but also on linking business investment to additional investments in education and research. The creation of the IIIT and ISB, for example, are essential programmes that ensure the long-term commitment of firms to the region as well as the capacity to develop the innovative and human resource base that will ensure the state's place in the global economy. Such investments are also complex alliances between the state, firms, local and international universities and non-resident Indians, guiding these institutions to become spaces for global best practice in innovation. As such, in both the growth of the regional industry and investments in the resources for future growth, the informational strategy can be viewed as a success.

5. The limits of software as an engine of growth

While it is clear that the information industry is well established and growing, the broader issue concerns the impact of the industry on the

Table 7.1 National and regional software exports (Rs million)

Nature of export	**1992–93**	**1993–94**	**1994–95**	**1995–96**	**1996–97**	**1997–98**	**1998–99**	**1999–2000**	**2000–01**	**2001–02**	**2002–03**	**2003–04**	**2004–05E**
National exports	6,750	10,200	15,350	25,200	39,000	64,000	109,400	171,500	283,500	317,220	461,000	628,120	875,000
Hyderabad exports	41.7	98.4	220	600	1,340	2,750	5,740	10,590	20,070	28,550	36,880	50,250	70,000
Hyderabad's % of national exports	0.62	0.96	1.43	2.38	3.44	4.30	5.25	6.17	7.08	9	8	8	8
Number of STP units	15	24	44	31	46	70	106	359	674	755	1,401	1,106	N/A

Source: NASSCOM at http://www.nasscom.org, accessed 15 December, 2004 and STP Hyderabad at http://www.stph.net/, accessed 15 December, 2004

Table 7.2 Direct technical manpower growth in Andhra Pradesh

Period	Technical staff	Non-technical staff	Total direct software
31 Mar 2001	53,870	4,130	58,000
31 Mar 2002	58,000	6,000	64,000
31 Mar 2003	64,295	7,150	75,000

Source: STP Hyderabad at http://www.stph.net/, accessed 15 December 2004

overall growth of the state. Clearly, the growth of the software industry within India has provided substantial growth in a leading technology sector for a country that had until the early 1990s missed out on all high-tech manufacturing industries. However, it has not provided the scale of growth to drive overall economic development in the country. Even after the extensive boom of the late 1990s, India's share of total global exports reached only 0.9 per cent in 2001, up from 0.6 per cent in 1995. This gave India a per capita export of US$60, six times lower than that of Brazil and three-and-a-half times lower than that of China (WTO, 2003).

On a regional level, these limits of software as a driver of growth were becoming evident even prior to the end of the 'perfect storm' conditions of 1999 and 2000. After having gained international recognition by attracting Microsoft to establish a software development centre in Hyderabad, Andhra Pradesh officials realised that overall growth was not being driven by such investments. As an example of this, after almost four years in Hyderabad, Microsoft's development centre employed approximately 100 individuals in 2002, an insignificant proportion of a state containing 80 million people. Higher levels of growth required much greater volumes of investment and employment generation. To meet these goals, the regional policy increasingly came to focus on ITES in preference to software or other higher value-added sectors.

The benefits to Andhra Pradesh of the policy shift were significant and immediate. Driven in large part by initial investments by GE in 1999 and 2000, as well as increasing investments by local firms seeking new revenue streams, ITES was the leading export sector from Andhra Pradesh by 2002. The second phase of the HI-TEC City technology park was almost exclusively dedicated to ITES. In contrast to the slow expansion of software services, a single major ITES or business process outsourcing (BPO) investment could generate up to 1000 jobs within a year of their establishment and millions of dollars in exports. By the end of 2002, Andhra Pradesh officials had established a specific governmental programme, APFirst, to facilitate ITES, and most leading software services firms have established ITES operations or divisions in the state.

The significance of this shift within the regional economy is not minor. Of the 60,000 new jobs created in Andhra Pradesh after 1999, 25,000 are in the IT-enabled services sector. These trends are also consistent at the national level, with three ITES jobs created for every two in the software sector since 1999 (NASSCOM, 2003) and the popularity of ITES clearly increasing.

6. ITES as a driver of regional growth?

The current emphasis on ITES raises two significant issues. The first of these is the difference between software services and ITES. The structure of software practice, specifically the demand for skilled labour and agglomeration to maximise productivity, limits the growth of offshore development. As indicated above, most of the leading Indian software services firms have extensive onsite operations on the global scale. Satyam, the leading Andhra Pradesh-based firm and one of the top five software firms in India, has seven global development centres outside India. Four of these are in the US and one each in Europe, Japan and China. It has also recently established an Asian regional headquarters in Singapore and made a significant investment in Malaysia's Multimedia Corridor in order to access local government contracts. It is also aggressively pursuing acquisitions to enter or strengthen access to new markets.

Traditional analytic frameworks from manufacturing like labour cost, infrastructure and rationalised work processes fit much better with understanding and predicting the direction of ITES than software services. On the positive side, ITES has the potential to be a much more general driver of growth, or at the very least to open growth to wider sections of the local population. As such, it has the potential to be a driver of middle-class growth that has been seen in China, Taiwan, Singapore and Malaysia – growth, however, driven by services and not by manufacturing. A single investment from GE or Citibank in a specific region may have the potential to have an overall impact similar to an Intel investment in Penang or Costa Rica, driving not only direct employment but also some growth in the local economy. The open question is what ITES services actually bring and require from a region, both in terms of a local network or production and innovation as well as the potential to upgrade. The ease of movement of ITES, the very basis for its sudden expansion in India away from rural areas in the US or the British Midlands, is an indication that such issues are not trivial.

This signals the second issue of longer-term growth, which raises questions concerning the negative aspects of the similarity to manufacturing. Software services – driven by domain knowledge, skilled labour and unique process-management skills – provide a competitive advantage to Indian firms that is difficult for other regions to emulate. ITES, in contrast, makes issues of infrastructure and labour cost much more central determinants of location, raising the overall competitive environment both in the near and longer term. While Andhra Pradesh, for example, has sought to distinguish itself from other regions within India, its move into ITES has immediately brought it directly into competition internationally with traditional strongholds of ITES such as the Philippines. In the even longer term, new states within India like Kerala – with good infrastructure and high rates of literacy – pose serious challenges to sustainable ITES sector growth in Andhra Pradesh.

7. The 2004 elections: a reality check on information capitalism

The National Democratic Alliance – comprising the BJP and key regional allies like the TDP – ran the 2004 election under the 'India Shining' plank, emphasising the emergence of India as an economically successful global power. In many respects, both the BJP and the TDP had room for confidence. They had presided over the emergence of India and Andhra Pradesh as key players within the global economy, centred not on low-cost manufacturing or agriculture exports, but on the most informational aspects of the new technology industry. In Andhra Pradesh, the government has led the state from an economically struggling region to a leading centre of software services and ITES within India and the world. These accomplishments were neither fleeting nor temporary, but involved serious investment and real growth. In spite of this, and in a completely unexpected outcome, the national alliance and the TDP lost dramatically at both the federal and state levels.

Why did this happen? The analysis above clearly indicates that growth has been significant and real within India and Andhra Pradesh. Even adjustments like the shift to ITES, while having distinct possibilities and directions for future evolutions, managed to maintain and expand economic growth during the period from 2000 to 2005. As such, how can the dramatic electoral shift be explained? Was it a referendum on the informational development policies or the extensiveness of the growth, or something altogether different?

The Congress Party's return to power in Andhra Pradesh was its first electoral victory in over a decade, and it is easy to assume that the election was a referendum on Naidu and the information economy-centred policies. However, all elections are over-determined, with a complex mix of factors determining the outcome (Virmani, 2004). In the case of Andhra Pradesh, it is too simplistic to say that the informational strategy was rejected or that the TDP lost because of its IT policy. A more accurate analysis recognises the complex mix of anti-incumbency bias, rising expectations, rising income inequalities, neglect of the large rural hinterland, and the overall economic structure's shaping of the outcome. This is equally true at the national level, where additional issues added to the electorate's calculations, including the repercussions of the riots and pogroms in Gujarat.

In Andhra Pradesh, the incumbency factor is the most easily accounted for. Naidu and the TDP were running for a third consecutive term, in a state that had rarely re-elected a government for even a second term. The success that enabled the party's original re-election is also part of the increasing expectations of the electorate. For the state overall and on a per capita basis, per capita income improved throughout the late 1990s and early 2000s, but it was unequally distributed. Most of the increase in wealth was captured by the urban middle class and the investors.

Even after the economic adjustments following the boom, the regional economy has continued to grow both generally and in key information industry sectors (see Table 7.3). The growth has not been dramatic, but it has been fairly consistent with national growth levels overall.

Table 7.3 Growth of GDP (constant 1993–94 Prices)

	1994–5 to 1999–00 (%)	**1999–00 to 2003–04 (%)**
India	6.20	5.60
Andhra Pradesh	5.60	5.50
India (Agriculture)	3.60	2.10
Andhra Pradesh (Agriculture)	3.50	2.63

Source: World Bank (2004)

What is clear, however, is that while the informational industries have shown growth and potential, the fundamental economic structure of Andhra Pradesh was not altered dramatically by the developments of the 1990s. In other words, the structure of the economy is still primarily rural and agricultural. Growth in software or ITES has done little to fundamentally alter this balance, and growth in these sectors, while important on a macro-statistical level, has limited impact at ground level outside urban areas. Throughout Naidu's second term in office, Andhra Pradesh suffered a combination of drought, a 25 per cent slowdown in agricultural growth and an overall deficit in the creation of rural employment. Such a combination, especially from the perspective of the rural electorate, made Andhra Pradesh 'a leader in reforms but not yet in growth' (World Bank, 2004, p. 15). This general sentiment could easily be applied at the national level as well.

Overall, then, the rural factor is an essential variable explaining the shift in electoral outcomes, especially when combined with rising expectations and the disappointment with Naidu in replicating the successes of his first term. In essence, Naidu and the TDP raised the bar and then failed to deliver, particularly to rural areas, in a time of crisis. These factors are even more significant given that 73 per cent of all population increases during Naidu's two terms were rural, representing an increase of over 7.2 million rural people. Given that surveys of rural residents on such issues as electricity, agriculture credit and other services show high levels of dissatisfaction (World Bank, 2003a), the emphasis on a modern, urban, tech-driven 'India Shining' was probably not the wisest election strategy in a democracy.

The point of the rural role in the election is to highlight distinct factors surrounding the impact of information industries within the larger economy. In reality, looking back to the 1990s, the state had very few options, and importantly the growth and sustainability of the industry justified the strategy. The real issue is why such growth and success did not translate into a more general transformation within the regional economy. The answer, in part, is that informational industries as they are structured at the time of writing do not have clear linkages with the rest of the economy, making general growth harder to promote and encourage. It is much more difficult to link software or services to agriculture than to other industries, even without the limitations of labour and skill so essential to software production. Even with a shift to the ITES sectors, the same limitations remain, particularly in relationship to the rural economy. And even where infrastructural or geographic barriers are weaker or even non-existent, the information-centred growth is much more about a better future than a better present. In other words, a call centre may serve a rural family's daughter well in the future, but it does nothing directly for a family trying to pay for the daughter to go to university or have constant electricity or water supplies at home in the present. This is one of the key lessons of the elections. It was not an invalidation of the informational strategy as such, but a questioning of how such growth is integrated and its benefits distributed for the overall social and economic development of a society.

8. The challenges of information-based development

The informational development push of Andhra Pradesh and India during the 1990s faced two very distinct tensions. First, there is the challenge of the informational policy institutionalising the inequalities that occurred while promoting the new growth strategy in the first place. Second, there is the open question of the software industry's contribution to overall growth. Both of these represent important issues to be worked through in light of the 2004 election results.

The first challenge of an ICT-led development strategy is that through the process of adapting specific national policies in support of any new initiatives the above dichotomies may become institutionalised. Even more significant is the possibility that such a development may expand the negative aspects of these legacies as resources and policies focus more on developing the innovative environment at any cost. Under these conditions, there is serious potential for a permanent digital divide (where access to technology, education and income are highly correlated), not between nations or regions, but between social classes, cities or even neighbourhoods, as inequalities surrounding education, quality of life, basic services and infrastructure are reinforced.

The second challenge for overall development represents a much more difficult issue. The overall social transformations linked to the general process of industrialisation – the rise of the middle class, urbanisation, consumerism, increased gender equity, expanded literacy, social movements – that have accompanied high-technology hardware industrialisation (Lubeck and Eischen, 1998; Rasiah and Best, 2000) have yet to be established as the default pattern for information-focused strategies. In contrast to previous hardware-centred strategies that many states had pursued, there is no clear trend or method that has demonstrated the capacities and policies states should develop to manage the challenges of a dynamic service industry. In other words, the process is only just being defined and navigated, both within India and globally.

Even the most established of these new regional centres – for example Bangalore, Karnataka – replicate the social and economic tensions present in Silicon Valley or other informational capitalist regions,[2] often starting from a significantly lower level of development and with much weaker mechanisms for social protection.

Strong and socially consensual policies are needed to embed such developments into the broader society rather than merely set up a parallel social and economic reality. If such policies do not exist, the risk is that the general spread of social and economic development will remain relatively stagnant, while new institutions and targeted investments link only small sections of the society to the global economy. As indicated in a very early assessment of Andhra Pradesh's IT strategy (Eischen, 2000), it was unlikely that export-focused and FDI-driven software could generate sufficient growth to address all the challenges of the state. Fifteen thousand highly skilled software professionals and 70 million dollars in exports were not sufficient to address the challenges of inequality in a population of 80 million with 17 million people living below the poverty line. While the industry has grown dramatically and there has been a significant shift towards ITES, the same challenges remain because of the nature of the informational economy. The ability of governments to generate policies that distribute and maximise the gains from information-focused strategies, while maintaining the organisational and social structures that provide those gains, is one of the greatest challenges of these initiatives.

The opportunity presented by software, and industries following informational patterns, is that local knowledge can be made use of and valorised within the production process itself. Regions should recognise that they are really developing a mastery of knowledge-based production processes, not software as an industry. In other words, focusing on synergy and developing the basic structures that promote an 'ecology of innovation' encompassing all sectors of the economy will produce economic and social development far outside of just software or services. It is the blending of software with other industries and domain knowledge that produces the real dynamism in the economy. Understanding this opens up the real possibilities for regional development strategies both to promote growth and to ensure its linkages with the entire economy. Existing industries and research centres, from media and film to agriculture and education, can become an integral part of a successful strategy, expanding the impact of the development process and ensuring its sustainability over time. All of this combined will establish the embedding of local knowledge into broader national and global processes, while lending the potential for the most negative social and economic divisions to be ameliorated.

9. Discussion and conclusion

Each of the aspects considered above for Andhra Pradesh is valid for India at large. As a second-mover with unique disadvantages, Andhra Pradesh provides a clearer picture of the issues involved, but such issues are not separated from the transformations occurring within India. The political factors that ended TDP's rule in Andhra Pradesh and BJP's control of the federal government also removed the incumbent Congress Party in Karnataka, the state that owns the crown jewel in the Indian IT industry Bangalore. While real growth has occurred and India and many Indian firms have set the global standard for offshore services, all states have faced challenges in linking this global success to widespread local development. It is also true that such success has increased the expectations of both urban and rural voters, anticipating not only economic growth but a new level of transparency and efficiency in governance. As such, the elections within India represent the outcomes of success as much as failure.

In evaluating the impact of this continued growth of the IT industry, the basic issue revolves around the ability of informational sectors to foster wider and deeper growth across the Indian economy. The 2004 elections indicate that such linkages are not naturally forthcoming. Policy changes that either ensure the benefits of growth are evenly distributed or foster linkages with the larger economy are necessary. Such linkages and the sustainability of these policies over time depend on the local adaptation of these resources to address both the challenges of the legacies and the impact of the informational strategy itself. Addressing these challenges means linking different knowledge industries, skill and markets to revitalise existing industries and develop new market opportunities. Institutionalising such synergy should be the goal of the overall development strategy. Such institutionalisation and the capturing of sectoral convergence involve social as well as economic transformations. In this way, viable strategies are about creating 'information societies', not just information industries.

The lessons from Andhra Pradesh suggest that the dynamics of the informational strategy and the potential for such synergies actually rest not necessarily in the industry or in a technology alone, but rather in the models of development that guide informational strategies. In other words, a central lesson of the past ten years up to the time of writing, in India is the inability of policy-makers, entrepreneurs and communities to envision and institutionalise general patterns of growth and its equitable distribution around informational industries. These are new models to be sure, and the elections of 2004 indicate that there is much to be learned about how to structure and ensure broad and equitable growth. However, the limitations, challenges and opportunities experienced through both software and ITES will be visible in biotech, multimedia and healthcare, to name just three emerging sectors. As such, if the lessons of 2004 can be properly understood, India will be well prepared to address the general dichotomies and opportunities of informational development.

What is essential to remember is that such challenges of informational industries are not limited to Andhra Pradesh or India in general. Many of the themes that resulted in such dramatic electoral changes within India could also be argued to have resulted in the election of Schwarzenegger in California. Real growth has occurred and will continue within India, but the structure and balance of informational development is still being worked through. The benefit of the elections is that India at the time of writing appears to be working through these issues. In essence, because the world is increasingly facing informational development questions and challenges, India is preparing for the future first, within a political structure of democracy that gives the country a self-correcting mechanism that other regions do not have in place and, as such, cannot benefit from.

This aspect should not be underemphasised, either within India or globally. The issue of development within India is not a technology problem or one that informational development can solve quickly or simply. The issue of development, just as it always has done, involves fundamental structural changes and institution-building.

Notes

1. While not huge when compared to VC funds in the US, Rs 150 million is large by Indian standards. The federal government VC fund designated for all of India, in comparison, was Rs 1000 million.
2. While beyond the scope of the analysis here, the social and economic patterns surrounding the software industry are clearly visible in other regions and industries, including Los Angeles (film), San Francisco (biotechnology and e-commerce), Austin (computers and ICs), New York (finance), Bombay (film) and Hong Kong (finance, e-commerce and film). The social implications of software clusters might be well informed by drawing on the careful analysis undertaken concerning these regional economies.

8
The Political Economy of the Indian Software Industry

Balaji Parthasarathy

1. Introduction

The Indian software industry has grown rapidly since the 1980s, with revenues increasing from less than US$100 million in 1985/86 to US$12.3 billion by 2002/2003. During this period exports grew even faster, with their share increasing from less than a third to more than three-quarters of revenues. This chapter explains the opportunities and constraints created by the increasing globalisation of the Indian software industry and the organisational changes that have accompanied the growth in software exports. Technological change over the period from 1970 to 2000 sustained the global demand for software that advanced industrial countries have been unable to meet. This created opportunities for new producers like India, thanks to the availability of a relatively large, low-wage, high-skill, English-speaking labour pool, and increasingly liberal economic policies. After entering the international division of labour by providing low-wage, low-skill onsite services, the Indian industry began to shift, increasingly, to more value-added offshore services in the 1990s to become the largest non-OECD software producer by 2000. Post-2001, the industry continued to grow despite a global slump in demand by capitalising, in part, on the globalisation of R&D services. Despite the transformation in the organisation and capabilities of the industry, the export of profitable software products and packages is limited. Besides explaining the differential successes in services and products, this chapter discusses the relevance of the case of India for other newly industrialising countries wanting to enter the global software industry.

Section 2 explains how the proliferation of PCs in the 1970s, networking in the 1980s and the Internet in the 1990s exacerbated the 'software bottleneck' that led to an explosion in global demand for high-skill, low-wage software professionals. Section 3 describes India's entry into the international division of labour of the software industry through 'body-shopping' and the subsequent shift to offshore services. Section 4 analyses the shift to R&D service provision by examining the changing technical and social divisions of labour in the production of embedded systems and the opportunities it opened up for India. Section 5 argues why India's success with services has not been matched in products and packages. The chapter concludes, in section 6 by highlighting the lessons that India's emergence as a software exporter offers to other newly industrialising countries (NICs) seeking to encourage the production and export of software.

2. The global context for Indian software production

An independent global software industry developed following IBM's decision in 1969 to sell software separately, instead of bundling it with hardware (Steinmueller, 1996). The decision was crucial, as it gave customers the option to buy their software and hardware from different vendors. The invention of the microprocessor, soon after, facilitated the development of powerful mini- and microcomputers that successfully challenged the dominance of IBM's larger and more expensive mainframes. With the availability of increasingly more powerful and inexpensive hardware, there was a proliferation of computer usage in various economic sectors, creating a huge demand for software.

While automated, capital-intensive operations permit the mass production of high-quality hardware, software production, in comparison, has remained a craft-like, labour-intensive affair, plagued by uneven productivity and quality and relying more on trial and error to achieve its goals (Gibbs, 1994). Software development was notoriously prone to 'bugs', delays and cost overruns. In other words, software productivity and quality have lagged behind those of hardware, creating a 'software bottleneck'.

To overcome the bottleneck, there has evolved a discipline of software engineering along the lines of industrial engineering. Managing the software development process, or structured programming, by breaking down the software development process into segments to permit co-ordination and control over a team of programmers working on a large project, has a long history. The six-stage 'waterfall' model – requirement analysis, specification, design, coding, testing and implementation – with each stage completed before flowing into the next is a classic conceptualisation of the software development process (Keuffel, 1991). Recent software management practices have focused on adopting industry-wide certification norms so that organisations adhere to codified quality procedures in the development process (Kitchenham and Pfleeger, 1996). Such norms include the ISO-9001/9000-3 standards prescribed by the International Standards Organisation and the Software Engineering Institute's five-level Capability Maturity Model (SEI-CMM).[1]

Although the widespread adoption of software engineering practices has helped to improve overall productivity and quality, the bottleneck persists due to wide variations in programmer productivity and quality (Gibbs, 1994). The extent to which process standards, such as those laid down by the ISO and SEI, will guarantee a quality product remains contentious (Schneidewind and Fenton, 1996). The waterfall model of development is criticised for enforcing a rigid top–down methodology on the grounds that no generic process model will work for all projects (Davis, 1993), much less one that is linear and sequential. In practice, software development tends to be iterative; there is learning by doing, with no sharp demarcation between conceptualisation and implementation (Keuffel, 1991). Thus, Kraft's (1977) prediction that structured programming would lead to a division of labour between mental and manual labour has not quite been realised.

Though the last three stages of the waterfall model have been amenable to automation with computer-aided software engineering tools, such automation demands tediously clear and detailed specifications. Consequently, much coding, testing and implementing is still done manually and the division of labour in software production is better seen as a division between more and less skilled labour. In short, quality and productivity improvements delivered by software engineering have thus far been limited, supporting Brooks's (1995) point that there is no 'silver bullet' for the bottleneck. Thus, overcoming the bottleneck has required the deployment of more software professionals.

Apart from the demand generated by the growing use of software, there is also plenty of older software still in use in need of maintenance. For instance, firms using computers extensively since the days when mainframes were dominant typically spend over 70 per cent of their systems budgets on maintenance activities because of the complex nature of the problem (Mayall, 1991). The hotchpotch structure makes maintenance and re-engineering labour-intensive.[2] Discarding older systems is often not easy, as they embody years of valuable business information and experience. One aspect of the hotchpotch structure is format problems, such as those involving dates, of which the Year 2000 (Y2K) problem received the widest publicity.[3]

The many such time-specific, labour-intensive demands coming together in the 1990s, in addition to the general expansion in the use of software, created a huge demand for skilled labour, especially in advanced industrial countries that used IT extensively. In the United States, for instance, though the question of whether or not there is a shortage of skilled labour is controversial, there is agreement that there is a shortage of people with specific skills (Koch, 1998). For instance, finding people familiar with old code written for mainframe systems is difficult. The shortage of people with specific skills and the pressure to limit the costs of labour-intensive projects forced firms in the US to look elsewhere for low-cost skilled labour. Among the countries able to offer this skill–cost combination was India, the top choice for onsite and offshore software development, ahead of Israel, Ireland, Mexico, Singapore, China, Hungary and the Philippines (InfoTech Consulting, 1992).

3. India in the international division of labour: from body-shopping to offshoring

The availability of necessary skill–cost advantage will not automatically guarantee the profitable exploitation of a global opportunity without a supportive socio-economic environment. It was for this reason, for instance, that the globalisation of the semiconductor industry since the 1960s had bypassed India (Henderson, 1989). Indeed, when India established the Department of Electronics (DoE)[4] in 1971 to formulate electronics policy, India's electronics production was larger than that of South Korea (Sridharan, 1996). But, a decade later, Korean production and exports dwarfed those of India. More broadly, India's share of world trade between 1950 and 1985 shrank from 1.8 per cent to 0.63 per cent,[5] as Indian economic policy between the 1950s and the mid-1980s was driven by a state-dominated, import-substitution industrialisation (ISI) strategy, with a commitment to self-sufficiency and self-reliance. Following IBM's forced departure from India in 1978, many of its former employees established service bureaus, which leased out computer time, before graduating to software development (Subramanian, 1992). This was the origin of the software industry in the private sector, although efforts to promote exports proved ineffective.

In the early 1980s, India took cautious steps to liberalise the policy regime to encourage private investment and trade (Sridharan, 1996). As a result, the 1980s witnessed many initiatives[6] that led to India's laying the foundation for a software industry (Lakha, 1990). While the conscious policies since 1984 were crucial in facilitating the change, reinforcing their efficacy were certain serendipitous benefits conferred by policies from earlier years. Despite widespread illiteracy, Indian education policies managed to create a large pool of skilled labour that, in a relatively slow-growing economy, remained a thoroughly under-utilised human resource. India's colonial legacy also meant that this labour was educated, mostly, in English.

India's most pointed advantage came, however, not merely from the availability of low-cost, English-speaking labour, but from the skills embodied in it. Following IBM's departure and the unsuccessful local efforts to build a commercially viable indigenous computer, users had to rely on imported computers. Since high duties were a disincentive to import, mainframes never made a significant presence in India, and the few that were imported were of various vintages and sources. As a result, in 1990 Computer Maintenance Corporation (CMC), the public sector firm established in 1975 with ex-IBM staff to maintain and service all computers not manufactured locally, was supporting 923 computers of nearly 60 models of various vintages made by 34 manufacturers (Harding, 1989; Subramanian, 1992). Indian programmers were working on a variety of platforms, which proved valuable in the 1980s in acquiring contracts for the maintenance of various legacy systems.

With the growth in computer manufacture and usage in India in the 1980s, Unix became the operating system of choice. As it undertook limited computerisation of some of its activities, the government also played an important role in encouraging the use of Unix. For instance, the Rangarajan committee's report on bank automation took the advice of CMC and recommended the adoption of Unix-based systems. This potentially large market encouraged investment in research and led to many innovations in the design of Unix-based hardware in the late 1980s and early 1990s (Heeks, 1996). Firms obtained the Unix source code and modified it to run on the machines they designed.

The importance of familiarity with Unix cannot be overemphasised. Unix itself was first developed by Ken Thompson and Dennis Ritchie at AT&T Bell Laboratories as a multi-user OS to provide a comfortable programming environment. Even as the usage of Unix spread, AT&T refused to support it, forcing users to come together to share information and ideas. AT&T's liberal licensing of Unix to many universities across the world was particularly conducive to fostering a collaborative spirit and Unix developed as a truly open system. The world's leading computer vendors adopted different versions of Unix (Salus, 1994), and it was under these circumstances that Indian programmers and engineers also adopted Unix (Udell, 1993).

Despite the expertise, exports grew slowly in the 1980s because Indian software firms were stuck in the relatively low value-added production segments such as coding and testing, or in maintenance and re-engineering tasks, as opposed to the more conceptually demanding and lucrative segments such as requirements analysis or design (Schware, 1992; Evans, 1995; Correa, 1996). Further, a substantial proportion of the low value-added software exports took the organisational form of onsite services that, in the case of Indian firms, meant 'body-shopping'. Body-shopping had its advantages and its limitations. It provided easy access to the global market, as it represented 'almost "inputless exports" which require only a contact overseas, a little finance, and the names of local programmers who can be hired if a contract is forthcoming. Such a path can also be followed relatively independent of government policy measures, particularly those on imports, because no imports are needed' (Heeks, 1996, p. 85).

The drawback, however, was that once sent overseas many employees tended to quit and seek better paying and professionally more challenging jobs. The high employee turnover only reinforced the tendency of Indian firms to compete on the basis of low costs rather than being able to fall back on a repository of technical and managerial expertise acquired from previous projects. It also led to an under-utilisation of Indian engineering talent.

Despite the drawbacks of body-shopping, and the kind of work that it entailed, firms had few other options. In the absence of easily available and affordable data communication facilities, firms could not shift to offshore services. Further, with the exception of a few firms such as Tata Consultancy Services, which has been an exporter since the 1970s, it was not until the 1980s that Indian firms began to make their presence felt as suppliers of software in the global market. Firms from a country that had hitherto merited no consideration as a source of IT products had to establish their credibility with global customers before the latter could be convinced that their software, which was strategically crucial to business operations, could be written several thousand miles away. A study of 230 projects across 125 firms in India by Banerjee and Duflo (2000) shows that reputation matters in software contracting, even after controlling for project, firm and client characteristics. A reputation could not be established overnight and had to wait until at least a few projects were completed satisfactorily. Thus, in order to gain the confidence of Western customers, there was no alternative to onsite services in the 1980s. Besides, while Indian engineers had the necessary technical skills, they had been trained in a different economic framework. Onsite services provided exposure to not just new technologies but also to management processes, market trends and socially specific communication protocols with customers.

The official encouragement given to body-shopping mostly reflected a limited understanding of the industry among policy-makers: software was widely perceived as being 'high-tech' without adequate distinction made between the different stages of production or the corresponding value-added.[7] Paving the way for a better understanding of the industry and policy support for local software development was a fundamental change in the approach to policy-making. Whereas until the 1980s it had been concentrated within a closed bureaucratic apparatus, since then the state has increasingly attempted to make policies by drawing on industry feedback (Evans, 1995). Based on input from the industry, a private sector industry promotion lobby, the National Association of Software Service Companies, or NASSCOM, was formed in 1988. A clearest instance of a proactive step at the public level was the establishment in 1990 of software technology parks (STPs). As export zones dedicated to the software industry, STPs offered data communication facilities with which firms could offer offshore services.[8] In 1991, the year after the STPs were established, a balance of payments crisis induced a major shift in economic policies, including devaluation of the Rupee, trade liberalisation and duty rationalisation, openness to foreign investment, and a new industrial policy that removed entry barriers to new firms.[9]

The shift to offshore services in a more liberal economic environment marked the beginning of a new relationship between the Indian software industry and the global market. Indeed, between 1985 and 2001 India's share of world exports increased from 0.5 per cent to 0.7 per cent,[10] helped by the rapid growth in software exports, whose share of Indian exports grew from 1.9 per cent in 1994/95 to 18 per cent in 2002/2003 (see Table 8.1). It is more than mere coincidence that the shift from a linear to an exponential trend in software exports that Sen (1994) points to began in 1993, the year that the first earth stations were commissioned at the Bangalore and Hyderabad STPs. STPs have also proven popular with industry, as the rising share of STPs in software exports illustrates (see Table 8.1).

STPs became popular because of the opportunities they provided in terms of the work done and value-added. For instance, firms could

Table 8.1 India's software and services revenues and exports, 1985/86 to 2002/2003

Year	Revenues (US$million)	Exports as % of revenues	Exports as % share of STPs	Exports as % of all exports
1985–86	81	29.63		
1986–87	108	36.11		
1987–88	130	40.00		
1988–89	160	41.88		
1989–90	197	50.76		
1990–91	243	52.78		
1991–92	304	53.95		
1992–93	388	57.99	8.0	
1993–94	560	58.93	12.0	
1994–95	787	62.13	16.0	1.9
1995–96	1,253	60.18	29.0	2.4
1996–97	1,841	59.75	46.0	3.2
1997–98	3,011	58.42	54.0	4.9
1998–99	4,069	63.90	58.0	7.6
1999–00	5,611	70.61	68.0	10.6
2000–01	8,386	74.14	70.7	13.8
2001–02	10,073	76.79	80.9	17.0
2002–03	12,324	79.29	80.7	18.0

Source: NASSCOM (1992, 1993) and from http://www.nasscom.org, accessed 20 February 2005. STP data from http://www.stpi.soft.net, accessed 20 February 2005

exploit the 12.5-hour difference between Indian Standard Time and US Pacific Standard Time to undertake maintenance and re-engineering after regular users in the US left work. This, in turn, meant lower costs and greater profitability, as overall resource costs in India are only about a third of what they are in the US. Offshore development also offered the advantage of having most employees under one roof, instead of their being scattered across many customer sites. This allows them to interact with one another and learn from other projects in a way that communication by e-mail alone does not permit. It also helped firms to build a repository of knowledge that helps in competing for subsequent projects. Though this did not overcome the problem of employee mobility and labour turnover, in the event of a crunch, employees could be moved from one project to another or new hands could be hired at local wages.

With the availability of data communications facilities, offshore software factories emerged complete with the infrastructure, technology, training programmes, quality processes, productivity tools and methodologies of the customer workplace. For instance, by October 1999 140 Indian firms had obtained the ISO 9000 certification offered by the International Standards Organisation, and 134 firms were awaiting certification (NASSCOM, 1999). By June 2002, 85 firms had obtained SEI-CMM Level 5 certification, compared with 42 in the rest of the world.[11] There are at least two reasons why quality certification is widely sought by the Indian software industry (Arora and Asundi, 1999). First, it is a marketing device, to signal to potential customers that the certified firm follows a well-defined and documented development process. Second, a well-defined process improves the ability of firms to estimate and manage the time and resources required for a project, helping them to bid for larger projects, thereby expanding business.

Although the relationship between certification and better rates is not very robust, for firms with an ongoing commitment to quality getting bigger projects is a route to obtaining more profitable contracts (Arora and Asundi, 1999). Firms with a proven record were able to bid for projects directly with clients rather than work as subcontractors for other vendors. This opened the doors to turnkey contracts, giving Indian firms responsibility of co-ordinating a much wider range of tasks than just programming. They had to take responsibility for the overall project schedule, quality and productivity. For firms that developed the managerial skills and proved their ability to deliver, there were significant benefits. Not only did some Indian firms get better work at better rates, but they also began to move away from competing solely on the basis of hour-based productivity to IPR-based productivity. They did so by converting critical knowledge gained during consulting projects, in specific application areas, to a generic product that is subsequently customised for clients with similar needs.

Even as Indian firms were establishing their credibility, many Indians who had migrated to the US were distinguishing themselves professionally. Since the passage of the Immigration Act of 1965 removed restrictions on Asian migrants, by allowing immigration based on either the possession of scarce skills or family ties to citizens and permanent residents, Indians went to the US for higher education and employment. Although Indians accounted for less than 5 per cent of all migrants during 1971–96, they accounted for nearly a fifth of all migrants with professional and special technical skills (see Table 8.2). The latter figure is more than double the norm for migrants from all other countries. By the 1980s, many Indians had entrenched themselves in positions of authority in various corporations and, by the 1990s,

Table 8.2 Migration from India to the United States, 1971–96 (As a percentage of all migrants and as a percentage of migrants with professional and technical skills)

Year	% of all migrants	% professional & tech skills (all migrants)	% professional & tech skills (Indians)
1971–80	3.65	N/A	19.50
1981–90	3.42	N/A	13.40
1992	3.52	7.57	20.98
1994	4.12	8.36	18.70
1996	4.68	8.22	17.94

Source: Alarcon (1998); INS, *Statistical Yearbook of the Immigration and Naturalization Service* (Washington, DC: US Government Printing Office, various years)

many were also taking the plunge as entrepreneurs.[12] On the basis of their own performance and the rising credibility of Indian firms, they could recommend India to their management, either as a possible software source or even as a production base in light of the policy changes taking place in India.

However, as with the case of Texas Instruments (described in the next section), when transnational corporations (TNCs) started trickling into India from the mid-1980s for the skills, they did not trust that labour with tasks that similarly skilled labour undertook in the home country. Many TNCs chose to put software engineers in India on a learning curve, especially when specific expertise was not readily available. Until such expertise was acquired, TNCs took advantage of the availability of programming skills before giving programmers the kind of responsibility that engineers in the home country were given. Some even started off doing internal body-shopping, that is, recruiting engineers in India and sending them off to do projects overseas with the parent company before starting their offshore development centre (ODC).

By the 1990s, however, with the establishment of the STPs and after the early entrants of the 1980s had demonstrated that India could be much more than a low-wage hunting-ground, there was a substantial increase in the ODCs of TNCs. Even IBM, which chose to leave India under the old policy regime, returned in 1992. Many ODCs shed their subordinate status and undertook projects either independently or jointly, as equal partners, with their parent organisations.

3. From turnkey contracts to R&D services

Following the slump in global demand after 2001, Indian firms attempted to take advantage of the efforts by firms worldwide to control costs by outsourcing not only software but everything from R&D services at one end of the skill spectrum to business processes, such as voice-based customer support centres (call-centres), at the other (Srinivas and Jayashankar, 2002). What is making the effort to provide R&D services possible is not just that India offers low cost options (since low costs alone cannot guarantee quality R&D), but also the availability of skilled manpower.

In the 1990s, the worldwide demand for skills drew Indian professionals in increasing numbers to lucrative opportunities overseas. Just one indicator of the phenomenon is that Indians became the largest recipient of H-1B visas to the US (see Table 8.3). This visa programme, established by the Immigration Act of 1990 to enable the employment, up to six years, of workers in 'specialty occupations', and used extensively by the US advanced technology sector, is widely identified as an 'Indian program' (Alarcon, 1998). But with the slump in the US since 2001, the number of H-1B applicants has gone down sharply, and NASSCOM estimates that approximately 35,000 professionals have returned to India, mostly from the US (Singh, 2003). Of the returnees, an estimated 70 per cent were H-1B visa-holders, while another 10 to 15 per cent has been abroad for at least ten years. These people are returning not only because they see international opportunities shrinking, but also because they see India as a place for innovation (Krishnadas, 2003a). One area where skill-intensive product development

Table 8.3 H-1B visas issued to Indian citizens and world rank

Year	Worldwide total	Visas to Indians (%)	Rank
1992	110,193	7.48	3
1994	105,889	16.01	1
1996	144,458	20.24	1
1998	240,947	25.96	1
2000	257,640	48.40	1
2001	330,521	48.90	1
2002	197,537	33.00	1

Source: INS, *Statistical Yearbook of the Immigration and Naturalization Service* (Washington, DC: US Government Printing Office, various years)

and outsourced R&D services are in evidence is embedded systems, a segment in which Indian firms and TNCs in India are very active.

An embedded system is any computer that is a component in a larger system and relies on its own microprocessor (Wolf, 2002). Although embedded systems have been around for many years, their use has grown with the increasing capability of microprocessors. They are now used in everything from consumer goods, such as cell phones and microwave ovens, to transport equipment, such as cars and aircraft, to industrial process control systems. An increasing number of devices use embedded systems to connect to the Internet and communicate with one another. In these devices, embedded systems take over what mechanical and dedicated electronic systems used to do. It is the ability to digitally capture and simulate various mechanical or other functionalities that makes the globalisation of R&D in various domains technologically feasible. Thus, for instance, firms such as Delphi, General Motors and Daimler Chrysler have set up R&D centres in India.

Embedded systems design is about adapting software abstractions, designed merely to transform data, to meet real-time constraints, power requirements and safety considerations in various domains while interacting with the physical world through sensors and actuators (Lee, 2000). Since the development of design algorithms for hardware–software co-design in the 1980s, embedded systems have been responsible for the 'softening' of once inflexible hardware, and hardware engineers at the time of writing this chapter, use methods similar to those of software engineers – they write tools and let programs generate the implementation (Vahid, 2003). Designing embedded systems often requires engineers who are classically trained in the domain of application, such as communications (Lee, 2000).

According to NASSCOM, the global embedded software solutions market was worth US$21 billion in 2003.[13] The telecommunications, computing and data communications segment dominated the market with a 34 per cent share, thanks to voice/data convergence and telecommunications deregulation in the US in 1996. The convergence required the design of new protocol stacks with open standards, as previously incompatible voice/data equipment of vertically integrated manufacturers had to communicate with one another. The technological and institutional transformation opened up opportunities for new firms, and this was the segment which Indian firms first entered (Bhuyan, 2002).

Facilitating the ability of Indian firms to offer their expertise globally in embedded systems, despite semiconductor-manufacturing capabilities in India continuing to be practically non-existent and logistical problems plaguing manufacturing, is the emergence of a new division of labour in the semiconductor industry. The vertically integrated firms that used to characterise the industry have given way to disintegrated firms occupying niches along a production continuum (Mathews and Cho, 2000). This continuum ranges from the pure play foundries at one end, which offer contract manufacture, to design houses at the other, which rely on other firms to manufacture the chips that they design. Firms in India have capitalised on the improved local ability to develop prototypes and taken advantage of the changed organisation of the industry by tying up with foundries elsewhere to produce their designs.

Firms were also helped by legal changes, important among which was the passage of the Semiconductor Integrated Circuits (IC) Layout-Design Act 2000 in the US, which provides for the registration and protection of IC Layouts and Designs for a ten-year period. However, while software patents are not permitted in India, on 27 December 2004 the Government of India issued an ordinance to modify the Patent Act 'to provide for patents when software has technical applications in industry in combination with hardware', that is, embedded systems, effective 1 January 2005.[14]

To take advantage of these opportunities, about 30 to 40 firms entered the business between 1999 and 2002, with investments ranging from US$0.5 million to US$2.5 million (Hari and Anand, 2002). By 2003 there were more than 100 firms (Hari, 2003), and it was estimated that the IC design and validation industry employed 5000 to 7000 engineers, compared to 300 in 1997 (Menon, 2003).[15] A different indicator of growth and size is the use of electronic design and automation (EDA) tools that are critical to IC design. The market for EDA tools is estimated to have grown from US$15 million in 1999 (Nair, 1999) to approximately US$60 million in 2003 (Menon, 2003). For instance, Cadence Design Systems, the largest EDA tools vendor in the world, which had only nine customers in India in 1997 (Nair, 1999), had 100 by 2002 (Hari and Anand, 2002). Of its 12 largest customers in the world, 11 have R&D facilities in India, including Intel, Motorola, TI, ST Microelectronics, National Semiconductor and Cypress Semiconductor (Kumar, 2003). However, due to the politically sensitive nature of the issue, firms do not necessarily publicise their expansion plans in India, which makes obtaining reliable data on the growth of the industry difficult. One indicator of the rudimentary state of the industry is the formation of the India Semiconductor Association (ISA) in October 2004 with 32 members to meet the specific needs of the emerging subsector of the software industry[16] as distinct from the more broadly based NASSCOM.

Although the data indicate that embedded systems production did not take off in India until the late 1990s, the origins of the industry can be traced back to the pool of IC designers employed by the Bangalore-based public sector firms Bharat Electronics and Indian Telephone Industries to meet their own needs (Sridharan, 1996). It was this pool of skills that TI capitalised on when it became the first MNC to establish an ODC in Bangalore in 1985, at the initiative of Mohan Rao, an Indian and a senior vice president with the firm. The subsequent evolution of the ODC provides an insight into the emergence of embedded systems production in India. Despite the availability of skills, TI did not immediately plunge into design. Instead, the Bangalore centre started off doing maintenance and application work, and it was not until the early 1990s that TI developed a design strategy in India.[17] A design centre for digital signal processors (DSPs), TI's main line of business, was set up in 1995 and Ankoor, the first commercial DSP fully developed in India, was launched in 1998. Since then, many other DSPs have emerged from the centre, in addition to 225 patents and a number of industry awards.

Chip vendors such as TI, however, do not develop their products in isolation. With the growing complexity of embedded systems and the rapid proliferation in their use, the industry is moving towards a design process that creates designs from scratch by integrating reconfigurable, commodity system-on-chip platforms to enable rapid design changes to offer differentiated products for a wide variety of users and application domains (Martin and Schirrmeister, 2002). Platforms are a mechanism for accelerating the design and development of end-user products by providing pre-integrated, pre-verified collections of IP blocks organised into hardware–software architectures.

If some of the bigger software service providers do not have the necessary background, a number of start-ups have emerged in various niche areas that would attract a chip vendor such as TI to India. The emergence is not only due to new market opportunities; more critically, it is the availability of the right kind of people, and because firm size *per se* does not offer any advantage as with services, which is encouraging entrepreneurs to take risks. In many cases the entrepreneurs are those returning home with not only years of design experience but also vital exposure to sales and marketing in the global arena. The lack of such experience previously inhibited home-grown start-ups and the interest of venture capitalists (Krishnadas, 2003b).

5. Software products and packaged software

In a study of the largest agglomeration of the software industry in Bangalore, Parthasarathy (2004) argues that, while the quantitative expansion in exports and the shift to higher value-added services reflects the ability of the Indian software industry to develop products for the global market, the limited contribution of software products and packages to Indian exports (see Table 8.4) reflects constraints on the ability to define new products and technologies. Specifically, the software industry in India has been unable to develop what Schware (1992) calls a 'walking on two legs' strategy to compete globally. The strategy entails creating a domestic market for various software domains to give firms an opportunity to develop their exports. The difficulty in pursuing such a strategy is reflected in the rising share of exports as a percentage of software revenues generated by firms in India (see Table 8.1). This, in turn, is an outcome of the relatively low spending on IT in India as a share of GDP (see Table 8.5).

To understand the role of the local market requires an explanation of the production of packaged software and its economic properties. The market for a package of functions opens up when there are enough users with similar needs, each of whom finds cost advantages in shifting from a custom product (Egan, 1997). Since the marginal cost of software production is very low in comparison to R&D costs, average costs decline with every additional user as the development costs of a package are spread across many users (Shapiro and Varian, 1999). For producers,

Table 8.4 Sources of India's software export revenues, 1990–2003 (as a percentage of export revenues)

Year	Onsite services	Offshore services	Products & packages
1990	90	5.00	5.00
1993–94	62.01	30.05	7.94
1994–95	60.90	29.59	9.51
1995–96	60.32	31.63	8.05
1996–97	58.69	30.21	11.10
1997–98	59.00	32.20	8.80
1998–99	58.18	33.91	7.91
1999–00	57.26	34.69	8.05
2000–01	56.09	38.62	5.29
2001–02	45.21	50.69	4.10
2002–03	38.95	57.90	3.15

Source: NASSCOM (2001, 2002)

Table 8.5 Information Technology spending and infrastructure, 1999

Country	IT as % of GDP	PCs per 1000 people
USA	4.14	517.07
OECD	3.09	285.55
Asia-Pacific	1.81	30.72
Malaysia	1.70	68.71
Taiwan	1.34	197.04
China	1.13	12.24
Philippines	0.78	16.92
Thailand	0.63	22.71
India	0.53	3.31
Indonesia	0.35	9.08

Source: ITU, Geneva at http://www.itu.int, accessed 20 February 2005

once a package is introduced for a specific set of functions, there are high entry barriers for a similar product. Software packages exhibit network externalities, where any increase in the total number of users increases the value of the package to the individual user, because establishing standards is important in the industry. Increasing returns to scale also favours monopolisation for any specific set of functions, since the price is inversely proportional to the number of customers and the lower the price the more attractive it becomes to a larger number of users (Egan, 1997).

To exploit the tendency towards monopolisation for any package of functions, first mover advantage becomes critical to producers, which in turn demands a good understanding of the evolving needs of users. To understand evolving user needs, however, access to markets is essential. It is here that the Indian industry is at a disadvantage. Although the availability of skilled labour at relatively low wages confers a huge advantage when it comes to development costs, the distance from final markets has meant that Indian firms are not able to react quickly with innovative products to meet the emerging needs of users. Despite the instantaneous access which modern data communication infrastructure provides to distant markets, the majority of Indian firms are stuck at the lower end of the business because the knowledge required to develop innovative products is often tacit and difficult to convey over long distances.[18] Evidence from the US packaged software industry shows how the local market matters. Over time, the industry has drifted gradually from regions where the computer manufacturing industry is dominant to regions with a concentration of the leading firms in various user industries (Egan, 1997). Thus, for instance, producers of software for the oil industry are concentrated around Houston, the centre of the US energy industry.

To overcome the limitations facing the software industry, and to take advantage of the internet revolution, the late 1990s witnessed many policy initiatives. In 1998, the Prime Minister established the National Task Force on IT and Software Development, with representatives from the government and the industry, to make recommendations to transform India into an 'IT superpower'.[19] Setting an export target of $50 billion by 2008 for software and IT-related services, the Task Force made 108 recommendations broadly relating to improving bandwidth availability and telecommunications infrastructure, fiscal incentives, and expanding IT usage to ensure 'IT for all by 2008'. In 2000, the government passed the Information Technology Act[20] to legally recognise electronic commerce and to facilitate electronic filing of documents with public agencies. In addition to national policies, in recent years at least 18 state governments have announced their own IT policies to attract IT investments and to encourage IT diffusion by focusing on e-governance, IT education and other related activities.[21]

Despite the lofty intentions, achieving the export goal for 2008 will present formidable challenges. A MIT report of 2000 indicated that, of the 108 recommendations made by the Task Force, 64 had been implemented, implementation was underway for 37, no action was taken on four, and the remaining three had been rejected.[22] As Verma (2000a, 2000b, 2000c) points out, the implementation of many of the Task Force recommendations has boosted the Indian IT industry, yet the dream of 'IT for all' remains very distant. This is because many of the recommendations implemented have been relatively easy, often within the purview of the governmental bureaucracy. Hampering widespread usage of IT are infrastructure deficiencies due to bandwidth constraints and unreliable power supply in much of the country, and low purchasing power as reflected in the low PC penetration in India (see Table 8.5). Further, there are the issues of low literacy and numeracy and limited availability of software in local languages.[23]

6. Discussion and conclusion

Rapid technological change over the period from 1970 to 2000 in information and communication technologies sustained the global demand for software that advanced industrial countries have been unable to meet. This created opportunities for new producers like India, thanks to the availability of a large, skilled, English-speaking labour pool and increasingly liberal economic policies.

While the political and institutional changes behind the liberal policies that allowed the Indian software industry to seize the market opportunities provided by technological developments since the 1970s were historically specific and socially contingent, the more than 400-fold growth in software exports in less than 20 years, nevertheless, offers a few lessons for other developing countries considering the software industry. First, the Indian strategy of entering the international division of labour by providing *custom* software services points to an alternative to the Taiwanese or South Korean model of *mass* manufacture that provided the means for their 'late industrialisation' efforts (Amsden, 1989; Evans, 1995).

Second, the Indian experience shows that, even with custom software provision, it is possible to enter the global market with the one key advantage that developing countries possess, low-wage labour (although for software the labour must also be relatively high-skilled) to transform one's position in the international division of labour on the basis of increasing interaction with global markets. While India entered the global market on the basis of low-cost, high-quality skills, organisational change and learning by doing have since led to an accumulation of capabilities. This was evident initially in the shift from body-shopping, where control over the productivity of employees and the quality of the output lies with the customer, to offshore services involving the acquisition of turnkey contracts and CMM/ISO quality certifications. As Indian software professionals and firms demonstrated technical expertise and increasing project management capabilities, it not only led to rapid growth in exports, but also began to gradually alter the relationship between the Indian industry and global markets.

Lately, the shift from software services to R&D services is being facilitated by a combination of technological innovation, organisational change and labour market changes. Technological change manifests itself with the blurring of the distinction between hardware and software, as it became possible to digitally appropriate and simulate various electrical and mechanical functions, and the growing importance of embedded systems in various devices. Organisationally, TNCs facing competition at home established R&D facilities in India. Major chip vendors, in particular, chose to rely on third party vendors to develop IP blocks and applications to sell complete solutions to a wide variety of consumers faced with ever-shortening product cycles. As global demand for IT fell, many highly skilled and experienced Indian expatriates followed the TNCs home. This widened and deepened the local labour market, raising the possibilities of bringing together various combinations of skills to make innovation and entrepreneurship more feasible.

The relative weakness with mass producing and exporting packaged software or software products leads to the third lesson from the Indian case. Critical to success in packaged software is the ability to embody intellectual content from various domains into code that is able to address the functional needs of users in a domain. The most effective way to continuously acquire evolving intellectual content in any domain is to locate close to the largest and most critical users whose decision to adopt any particular software as standard typically influences other users and unleashes network externalities. But India's failure to pursue a 'walking on two legs' strategy, which pays as much attention to policies encouraging domestic demand and consumption of software as it does to supply and production for export, has hindered the domestic market from providing a learning opportunity before venturing into the more competitive global market. Considering that the global market for packaged software grew at 11.8 per cent in comparison to 7.8 per cent for software services between 1990 and 1997 (OECD, 2000a), and that the packaged software business is more profitable than services due to the near-zero marginal cost of electronic duplication, India's relative failure in nurturing a domestic market offers another lesson to ponder.

Notes

1. SEI's five-stage model was originally developed to assist the US Department of Defence in software acquisition (Paulk, 1995).
2. Re-engineering is a means of reducing and simplifying maintenance by essentially replacing the older software with equivalent new software, that is, changing its form without altering its function (Sneed, 1995).
3. Jones (1998) points to other format problems with older software that manifested themselves in the 1990s.
4. Renamed the Department of Information Technology since November 2000. In this chapter, DoE and DIT will be used interchangeably.
5. Available at http://indiabudget.nic.in/es96-97/CHAP6.HTM, accessed 20 February 2005.
6. Two key initiatives were the Computer Policy of November 1984 and the Computer Software Export, Development and Training Policy of December 1986 (Subramanian, 1992). The 1984 policy eased the local manufacture and

availability of computers, and recognised software as an 'industry', making it eligible for investment allowance and other incentives. It also lowered duties on software imports, and made software exports a priority. The 1986 policy aimed at increasing India's share of world software production. The industry was to be independent, with the government stepping in to provide only promotional and infrastructure support. Overall, this policy was a rejection of Indian ISI and the ideology of self-reliance in the software sector.

7. Interview with N. Seshagiri, former Additional Secretary, DIT (24 June 1996), and N. Vittal, former Secretary, DIT (25 June 1996).
8. For details of all the benefits offered by STPs, see http://www.stpi.soft.net/stp_scheme.html, accessed 20 February 2005.
9. For details of these policy changes, see Acharya (2002) and the essays in Oman (1996).
10. See http://indiabudget.nic.in/es2003-04/chapt2004/tab75.pdf, accessed 20 February 2005.
11. Available at http://www.nasscom.org/artdisplay.asp?cat_id=205, accessed 20 February 2005. Although SEI upgraded the CMM model to CMMi (Capability Maturity Model Integration) in 2000, the broad philosophy of the five-stage model remains the same. For details, see http://www.sei.cmu.edu/cmm/cmm.html, accessed 20 February 2005.
12. See Saxenian (1999), Saxenian et al. (2002) and Alarcon (1998) for a study of the impact of changes in US immigration laws and the role that immigrant engineers play in the advanced technology sectors of Silicon Valley, arguably *the* most important centre for the US IT industry. The studies also show the importance of immigrants for entrepreneurship in the Valley and how immigrant engineers use their social networks to promote investment and economic development in their home countries.
13. See http://www.nasscom.org/artdisplay.asp?cat_id=615, accessed 20 February 2005.
14. Available at http://pib.nic.in/release/rel_print_page1.asp?relid=6074, accessed 20 February 2005.
15. According to Singh (2003), one industry representative believes that one in three engineers joining the chip industry is a returnee from the US.
16. For details, see http://www.indiasemiconductorassociation.org, accessed 20 February 2005.
17. Unless otherwise mentioned, details of TI's activities are from a personal discussion that followed a presentation by Biswadip Mitra, Managing Director, TI India, at the Indian Institute of Information Technology, Bangalore on 10 April 2004, and from http://www.ti.com/asia/docs/india, accessed 20 February 2005.
18. Arora and Athreye (2002) raise another aspect of this issue by suggesting that Indian firms have faced a chicken-and-egg situation when translating domestic experience into global opportunities. They argue that, even if the Indian market were large, both in absolute terms and in terms of the range of learning experiences it afforded firms, it may not be relevant for export markets, especially to the extent that global customers are reluctant to contract projects which may force them to share the domain expertise that Indian firms lack. This, in turn, limits the incentives for Indian firms to use the domestic market as a global launch pad.

19. For details of the Task Force constitution and its recommendations, see http://it-taskforce.nic.in/, accessed 20 February 2005.
20. Details of the Act can be found at http://indiacode.nic.in/, accessed 20 February 2005.
21. The policies of most provinces can be found at http://www.nasscom.org/artdisplay.asp?Art_id=13, accessed 20 February 2005.
22. Report available at http://www.mit.gov.in/atrnt.htm, accessed 20 February 2005.
23. Studies conducted by the consulting firm McKinsey for NASSCOM also reflect these problems. In a 1999 study, it projected that India could export $30 billion of IT services by 2008, and estimated the domestic IT market at $20 billion. In a 2002 study, however, McKinsey changed its projections for the domestic market in 2008 to $15 billion, even as its 1999 export projections remained unchanged (Deccan Herald, 2002).

9

Working with Uncertainty in the IT Industry

Biao Xiang

1. Introduction

Srinivas, a 23-year-old IT professional from South India, went to San Francisco in 1999 through a placement agent. In 2001, the first of his three flatmates (all were Indian IT professionals) left for India after being jobless for three months; the second lost his job but moved to Canada to try his luck; and the third moved into a cheaper flat with his cousin to save money because he was expecting to be laid off. Srinivas started performing *puja*, a Hindu ritual, every Friday and Monday morning in early 2001, because IT firms in the United States usually announce their lay-offs on these two days. Despite the long *puja*s, he received the proverbial e-mail asking him to go and subsequently returned to his nominal employer, an Indian-run small placement agent for new job openings. Eventually, Srinivas decided to return to India in June 2001. His father had just started bride-hunting for him, projecting his son as a 'US-based IT professional', the top status in the Indian marriage market. Srinivas's return spoiled all his father's plans, and his father asked me why George Bush had done this to him.

Srinivas's story is just one of many – some far more dramatic – that are related by his fellow IT workers. The story brings into question the popular image of the IT professional and, more importantly, challenges the widespread myths that knowledge, possessed by workers, would generate rapid wealth and that labour relation would no longer be an issue in the so-called IT-driven 'New Economy'. On the contrary, Srinivas's experience shows that labour is not only an issue, but it may be more salient in the IT industry than in other sectors due to the high level of volatility and uncertainty experienced by the industry. It was predicted that the year 2001 would experience a shortage of 850,000 skilled professionals in the IT sector in the US alone (see Steen, 2001). However, more than 350,000 high-technology jobs were lost in the first eight months (Dunn, 2001) and lay-offs amounted to 600,000 by November (Shiver, 2001). Under these circumstances it is important to explore how labour fares in general and how it is managed, in particular, in the IT industry.

This chapter addresses this much neglected but crucial aspect of the globalisation of labour and capital, based on my fieldwork among migrant Indian IT workers. The chapter demonstrates that central to labour management in the IT industry, at least in the present case, are three processes. First, there is the process whereby IT companies outsource labour management tasks to placement agents, which enables the IT companies to respond to the volatile financial markets and ever-changing technology trends as flexibly as possible. Second, placement agents exercise direct control over labour on a day-to-day basis through the process of interaction with them. Finally, there is the process whereby individual workers move up the career ladder within the existing institutional system, which in turn enhances the labour control mechanisms. This chapter, as an ethnographic configuration of the labour relation, will not only detail these processes, but also demonstrate the ways on which they are interlinked and make the labour control system self-perpetuating. In its analytical approach, this chapter intends to bridge the levels of structure and agency on the one hand, and integrate institutional analysis and enquiry concerning subjectivity issues such as ideologies and perceptions on the other.

India has become the major source of migrant IT labour worldwide, with Indians having accounted for 63 per cent of all foreign IT professionals entering the United Kingdom in 2000 (Khadria, 2001). In the US, of all the computer systems analysts and programmers on H-1B visas, a visa type created in the early 1990s specifically for high-tech personnel, 74 per cent had been born in India (INS, 2000). Many Indian IT workers were managed through an almost unique system known as 'body-shopping' in the industry, which embodies various characteristics of the general global IT labour market. These two aspects are inherently related, for it is partly because of the highly effective labour control mechanism that Indians became particularly competitive, taking a lion's share of the global IT employment market.

In what follows, I will first construct a background for the core discussion by examining the developments in the IT industry at the time of writing, highlighting its volatility and uncertainty. I suggest that the IT industry achieved phenomenal growth in the 1990s due to its connection to global finance capital, and this connection made a high level of volatility an inevitable part of its dynamic. Then I delineate the historical evolution and the present structure of the labour management outsourcing system, which forms the basic macro-institutional setup of labour control in the market. The fourth part of the chapter explores how workers are directly controlled on a day-to-day basis, through a description of the internal structure of small agents. After detailing these control mechanisms, it is noted that young, educated and ambitious IT workers are not just passive objects, but they also constantly seek out new opportunities because of their unique cognitive capital. Finally, I focus on workers' perceptions and strategies and clarify the relations of these strategies to the larger institutional arrangement.

The chapter is based on 22 months' fieldwork by the author, in Sydney (from January 2000 to June 2001), Andhra Pradesh (from June to September 2001, mainly in the capital city Hyderabad) and New Delhi (from September to October 2001). In-depth interviews and participant observation comprised the main investigative methods. The author had open-ended interviews with 180 IT professionals and relevant institutes and lived with about 20 Indian IT workers in Sydney and in Hyderabad. Besides first-hand data, I accumulated considerable information through documentary study, particularly by collecting media reports. In order to maintain privacy and confidentiality, most of the individual interviewees' names mentioned are in fact pseudonyms. It is not claimed that my informants' experience captures the entire gamut of that of migrant IT workers from India. Rather, what this chapter claims is that labour control is an important aspect of the political economy of information capitalism.

2. Information technology and global capital

Although the IT industry came into being in the 1960s, it was only in the late 1990s, when IT was applied widely in various businesses in the form of application software and the Internet expanded rapidly to meet demands for content delivery and e-commerce, that it became a major growth industry. Along with such rapid growth came volatility. These two changes are, in fact, two sides of the same coin: they are the result of the industry's close connection to global finance capital. IT functioned as a crucial tool for the 'flexible accumulation' (Harvey, 1989) of global capital. Business-integrated IT made 'deterritorialised' production and management technologically practical and possible. E-commerce was arguably leading to a 'tariff-free' economy and would enable capital to pursue profits globally through free trade regardless of its origin and size. Internet technologies enable firms to manage resources on a global scale. For example, through 'B2C' and 'B2B',[1] consumers, material suppliers, manufacturers and dealers, wherever they are, can all be linked together in real time in order to reach the optimal arrangement of production and management. It was expected that with Internet technologies the true spirit of capital to maximise itself by circulating freely could be finally realised.

At the same time, IT became the major arena where global capital was concentrated, particularly through the stock market such as the NASDAQ. Having started to trade in 1971, NASDAQ surpassed the New York Stock Exchange in the mid-1990s to become the largest stock exchange in the US. Dominated by high-technology firms, NASDAQ is also the first fully 'electronicalised' stock exchange and literally absorbs money from all over the world. NASDAQ transmits real-time quote and trade data to more than 1.3 million users in 83 countries (NASDAQ, 2001). In terms of the source of the capital, the most globalised fraction of capital – finance capital – was the first major commercial investor in the IT sector and is also a critical consumer of IT products. According to a McKinsey Report (cited in Jain, 2002), banks in the US invested more than US$5000 per employee in IT, or around 12 times the investment of the rest of the private sector. Software for the banking sector was the focus of most Indian software developers in the early 1990s (*Financial Times*, 1995). Other established global companies such as airlines are also major clients of the IT industry.

The IT sector also became a major generator of global capital throughout the second half of the 1990s. California has overtaken Italy and France to become the world's fifth largest economy, mainly due to its IT industry (*Computer Today*, 2001). Bill Gates's personal wealth in 1998 amounted to more than the GNPs of 85 nations. The rosy expectations surrounding IT attracted further investment, much of which was speculative rather than based on a clear understanding of how the profit would be realised through concrete outputs.

The industry's close connections to global capital, however, also brought it unprecedented volatility. In March 2001, gloomy profit reports from Cisco System and Microsoft created global panic in the IT sector. Share prices on NASDAQ plummeted by almost a half compared to its early 2000 level. The so-called 'dot-com' companies, who had been the central players of the IT boom, suffered the most casualties. Rather than producing any tangible products, dot-com companies created websites and provided certain Internet-based services to attract visitors, and then used the visitors as the leverage to sell the websites to big companies. For example, Hotmail, set up by Sabeer Bhatia, an Indian software engineer in the US, was sold to Microsoft for US$400 million. Junglee, an Internet browser founded by four Indians, was acquired by amazon.com for US$180 million. Dot-com companies were normally supported by venture capital and had no cash flows of their own. They typically lasted from six months to two years before they were either bought by large companies or closed down. When the panic in the stock market spread, venture capitalists abruptly ceased investing. Venture capital deals in IT in the second quarter of 2001 totalled US$3.14 billion, down from US$9.28 billion the year before (*Times of India*, 2001a). Three hundred and thirty dot-com companies closed down in the US during the first half of 2001 (Nag, 2001).

Due to the particular nature of the dot-com and other IT business models, these 'companies' needed a large number of temporary contract workers, particularly those with Java skills.[2] The slowdown immediately created massive lay-offs. One estimate suggested that 35,000 dot-com workers had lost their jobs between January and March 2001 in the US alone (Vaas, 2001). More strikingly, according to Challenger, Gray & Christmas, a placement firm in Chicago, dot-com companies announced 98,522 layoffs in November 2001, more than doubling the 41,515 firings in 2000 (*Times of India*, 2001b). Therefore, the uncertainty in the labour market is an intrinsic part of the IT industry and is directly attributable to globalisation.

3. IT placement business: macro system of labour control

In order to understand how labour is managed and controlled in the extremely volatile IT industry, we need first to grasp the basic structure of the global IT labour market. Far from resembling the neoclassical model of the labour market where workers bargain with employers, the global IT labour market is dominated by a special placement business and the supply and demand sides of labour interact only through intermediate agents. In Australia, there were about 4000 agents involved in IT placement (Mulcaster, 2001) and 50 per cent of all IT companies were estimated to rely on placement agents in labour management.[3] Most IT job advertisements in *The Australian IT* and *Fairfax*, the two main media sources for IT job information, were posted by agents rather than employers. In the US, there were hundreds of agents specialising in H-1B visas for IT workers in Northern California alone, and perhaps over one thousand across the country in the year 2000 (Lubman, 2000).

Among Microsoft's 20,000 workers at its headquarters in Redmond, Washington, 6000 were employed not directly by Microsoft, but by temporary employment agents. Many workers had come to be known as 'perma-temps' because they had worked for Microsoft for years but were still managed by agents (Borger, 2000). Some informants suggested that both in Australia and the US it was big placement agents, rather than IT companies, who determined the labour price at the market.

Placement became a significant business worldwide in the 1970s as part of the 'flexibilisation' process in the labour market, particularly in semi-skilled service sectors, such as tourism and secretarial services. Labour placement agents, also called body/labour-hiring or body/head-hunting agents, were essential to mediate the flexible labour market. IT placement is part of this business, but has its own distinctive operational patterns. This can be illustrated by an overview of the historical evolution of the business in Australia.

3.1 IT placement business

The emergence of the IT placement business in Australia can be traced to a peculiar scenario in the early 1980s. A group of entrepreneurs who started with speculations in real estate invested heavily in media, insurance and the then-burgeoning IT industries. This created a new demand for IT professionals and placement agents brought a group of workers from overseas, mainly from Europe. This was possibly one of the earliest IT labour placement deals. The wide application of IT to various industries in the 1980s, mainly in the form of software, made IT expertise extremely valuable. A new employment pattern, namely 'contract consultant', came into existence. In this pattern, an IT professional registered a company under his or her own name, but typically did not hire anyone else or conduct business independently. The IT professional would use the name of the company to sign contracts with other companies (clients) to assign him or her to work for the client. A contract consultant's income was usually 2–2.5 times higher than a permanent employee's, since they were not entitled to superannuation and fringe benefits. At the same time, employers also preferred project-based contract workers to permanent employees due to the ever-shortening life cycles of the IT technologies. This employment pattern gave a new boost to the placement business, since both contract consultants and employers needed agents to match them up.

During 1990/92, the entire capitalist world was in the midst of an economic recession and Australia was no exception. A large number of unemployed workers flooded the employment market. Some workers of traditional professions, such as mechanical engineers, shifted to the growing computer industry. Contracting was a typical way for them to enter the new sector and obtain basic hands-on experiences. On the other hand, the recession made employers highly wary of overhead costs of permanent staff, preferring to make do with contract workers who could be retrenched at any time.

By the end of the 1990s, due to the so-called Y2K problems and the 'dot-com' boom, contract employment became a norm in the IT industry. It was estimated that more than 30 per cent of IT professionals worked on a contract basis in Australia at the turn of the century.[4] According to a report by Deloitte Touche Tohmatsu, a major consultancy company, even senior managers were increasingly working as contractors (cited in Cresswell, 2000). Contract jobs are normally managed through placement agents.

Apart from the demand for a flexible labour market, placement agents are also needed for institutional reasons. In Australia, employers were subject to complex labour regulations, in particular the Awards System. According to these regulations, a court could judge the nature of the relationship between an employer and an employee by substance rather than form, and therefore may deem a long-term temporary contract job to be one of permanent employment, notwithstanding the existence of a contract (Nolan, 2001). I was told that it often happened that an employer could not terminate a contract with a temporary worker because of this. When a placement agent is involved, the employer has no direct legal relationship with the worker and therefore is free from such liabilities. In this sense, placement agents were essential for establishing an institutionalised flexible labour market.

The IT placement agents' institutional function is even more evident in international recruitments. Because of time–space constraints and high costs, most employers were almost completely dependent on international recruiters in need of foreign talents. Companies often saw sponsoring overseas workers troublesome due to various government regulations. For example, according to the regulation concerning the Australian 457 visa scheme, a long-term temporary employment visa that most Indian IT professionals used to enter Australia, sponsors were obliged to assume financial responsibility for foreign workers so as to limit the social and financial costs borne by the Australian taxpayers. The sponsor had to bear a liability of A$2000 per month for each worker, including A$1000 for accommodation, A$200 for the telephone bill, A$500 for food and so on. The situation in the US was similar. In order to sponsor an H-1B visa, a company had to go through a procedure with at least four government agencies, namely the local employment service, the Department of Labour, the Immigration and Naturalisation Service[5], and the State Department (Kaplan, 2000). The sponsor was also responsible for compliance with the worker's conditions of entry as well as with the provisions of the labour law. But the increasingly uncertain market situation made it difficult for companies to promise on-time payment and working conditions in advance. Furthermore, various forms of co-operation between companies, such as alliances and joint development, which have become increasingly common in the high-tech industry, made it difficult even to identify whose responsibility it should be to sponsor a worker for a joint project.

Finally, while a 457 visa in Australia was valid for 1 to 4 years, many IT projects lasted only six months or less. Very few companies were willing to be responsible for a foreign worker after the project finished. At the same time, companies were not happy to go through all the administrative procedures to sponsor each worker for every short-term project.

These frictions between the state and the market necessitated third-party entry. By specialising in placement and migration activities, placement agents could deal with these problems much more efficiently. For example, agents could manoeuvre between different types of visas such as tourist, short-term business and student visas to get the worker into a country quickly, and transfer their visa to a working visa afterwards. When Uday, my key informant and a 29-year-old IT worker from India, met a placement agent in Singapore, his social visit visa was due to expire in 14 hours. The agent arranged another short-term visit visa for him within two hours and sent him to work the very next day, and the rest of the procedure was dealt with in a few days.

These institutional backgrounds result in the key feature of IT placement agents, namely, unlike the conventional placement agents who *introduce* employees to employers, the IT placement agents *manage* workers for the employers. Workers managed by IT placement agents do not form formal employment relationships with IT companies, but they remain employees of the agents. In many cases agents project themselves as technology firms, and their provision of labour to clients appears in document as an inter-firm transfer, that is, sending workers to a partner company to conduct a certain task. This is similar to temporary help agents, who also employ workers and send them out to customers to work on an hourly basis at the client's premises and direction (Kalleberg, 2000). But apart from their different market niches (semi-skilled versus highly skilled) and scale (local versus global), IT placement agencies are also special in that they often form 'agent chains' and adopt a practice of 'benching'.

3. Agent chains, body-shops and benching

Agent chain refers to the phenomenon where a group of placement agents of different sizes depends on and connects to each other in a business. In the market, major IT companies normally outsource their comprehensive labour management tasks, including recruitment, billing, payment and, even, employee promotion, to one or a few agents who fully understand their overall business, and only large agents have access to the major companies. But it was not economical for large agents to recruit and manage a large number of individual workers, particularly in a situation where the client needs workers with specific skills on short notice. For this reason, they turned to medium-sized agents for employees. Medium-sized agents recruited workers themselves, but some further relied on even smaller agents to provide workers to them. Small agents, as I will demonstrate shortly, had the advantage in recruiting and managing a handful of workers at one time by utilising their networks and by circumventing some regulations. Small and, in many cases, medium-sized agents could not approach major clients. In Sydney, the biggest five to seven agents in the IT area were estimated to take 40–60 per cent of the market share, and the smaller ones had to rely on big agents to access the market. Big agents did not contact small agents directly because of their lack of trust in the small agents and also because it is inefficient to deal with a large number of small agents, hence the need for medium-sized agents.

How do agent chains work? A small agent sponsors a worker, then loans the worker out to a medium-sized agent, who is then sent to a bigger agent and finally placed to a client where the worker does the real work. I came across three Indian workers who were placed to a client through five agents. In this situation, the client signed a contract with only the last big agent. The client paid the worker's remuneration as 'service fees' to the big agent, as though the agent assigned its own employees to work for the client. Big agents usually take a cut of A$10–15 per hour, out of the total payment of A$30–100. Subsequently, the big agent forwarded the pay to the medium-sized agent, still in the name of a service fee. The amount of margin typically cut by a medium-sized consultancy is 20–30 per cent of the workers' pre-tax remuneration. Finally, small agents took a flexible margin from the worker's wage. From workers with one to three years of experience, the agent might take a margin as high as 50–60 per cent before tax. In one case, a worker was paid A$6800 per month by the client, but received only A$1200 in the end. For workers with more than five years of work experience, the margin may be lowered to 20 per cent.

In the Indian context, the small agents who act as overseas workers' sponsors and legal employers are called 'body-shops', and related to this the practice of recruiting IT workers from India and then farming them out is dubbed 'body-shopping'. According to my field data, there was a total of 35–60 Indian body-shops in Sydney by late 2000, managing some 1500–2000 contract workers.

Indian body-shops assumed two crucial functions for the agent chain and the IT placement business in general. The first is to recruit workers internationally and to create a labour pool. Body-shops normally found workers through 'friends of friends', which made their recruitment practice highly efficient. When Kana, the owner of the Sydney-based body-shop Advance Institute, went to Singapore to recruit workers, he sent e-mail messages to their friends explaining what kinds of people they were looking for. Their friends subsequently forwarded these messages to their fellow Indian workers. Within two days after arriving in Singapore, Kana interviewed 20 workers. Four days later, six workers were selected. Kana also encouraged workers whom he was sponsoring or had sponsored to bring more people when they were in India for home visit. About five workers at the Advance Institute had been recruited in this way. Body-shops also recruited workers through overseas associates, mainly their counterparts in India. Body-shops in India arranged for IT professionals to go overseas mainly by asking their associated body-shops in the destination countries to offer the sponsorship. Sometimes the overseas body-shop paid the Indian agent, and in other cases the reverse occurred, depending on the market situation and the workers' skills. The body-shop in India charged a high fee from the worker as its profit and the agent overseas made money mainly by cutting a margin from the worker's pay.

It is important to note that the Indian body-shopping system was basically supply-side driven, that is, body-shops brought a worker from India overseas mainly because the worker was willing to go and, more importantly, willing to pay, rather than satisfy a definite demand in the destination country for that worker. The body-shops searched for job opportunities for the workers only after they arrived. Body-shops had to do this because without workers available for job interviews immediately, no bigger agents would talk to them. In other words, body-shops made bigger agents rely on them mainly by creating a labour pool ready for use, thus saving big agents and companies from going through all the administrative procedures needed for recruiting workers from overseas and enabling big agents to address urgent demands from clients at any time.

However, this created a problem for the body-shop, namely that it had to constantly manage workers without jobs, and this led to another practice peculiar to body-shops, 'benching'. This means sponsoring workers to move to a new country without any job opportunity in existence, with the result that the workers had to wait 'on the bench' after they arrived. Bigger agents normally did not do this since technically it was illegal, and controlling workers on the bench was the second key function that body-shops assumed in the agent chain. Originally, the word 'bench' in the industry meant the interim period occurring after one contract has finished and before the next one starts, during which time the worker has to wait on the 'bench' for jobs. Since bigger agents turned to body-shops for workers only when they had definite job openings, body-shops had to manage workers on the bench between projects as well, in addition to the new arrivals. During the bench period, workers are not paid or are only given a nominal 'bench salary'.

The agent chain and benching practice created a reserve army from which corporations could select and dispose of workers at any time, yet without producing costs to the state as welfare recipients or to the firms as pensioners. As an Indian H-1B worker in the United States put it, the body-shopping scheme is similar to the way airlines overbook flights, to have workers available in case a need arises (Kirby, 2001). One of my informants put it in this way: 'The companies need a big reserve [army] of workers. But it is very expensive to keep all the workers for one big agent. Then these body-shops came in.' The question now becomes, how do body-shops manage the workers on the bench?

4. Internal structure of body-shops: micro labour control mechanisms

When undertaking benching, body-shops had to handle two thorny issues. First, they must prevent workers from leaving them soon after arrival. Bringing a worker from India to Australia could cost as much as A$10,000[6] and this could not be paid off unless the worker worked for the body-shop for at least six months. However, it was certainly unpleasant for a worker to be on the bench, particularly when there was an acute demand for IT professionals in the market. Furthermore, the Australian regulations with regard to the 457 visa allowed workers to change their sponsorships without permission from their sponsors. Nor would the change impede their future applications for permanent residency (PR), which was the case for the H-1B visa holders in the US. This means, at least in theory, that the worker could jump ship at any time.

The second thorny issue was that workers may raise complaints to the government against the body-shops, and such complaints could lead to severe punishment by the authorities. When applying for sponsorships, body-shops usually made generous offers to workers in the documents that they submitted to the immigration authority in order to go through the administrative procedures smoothly. This was done not only to meet the minimum wage level designated by Australian law, but also to show that the person sponsored was senior and could contribute to upgrading local skills, which is a type of immigrant strongly favoured by the government.

However, workers rarely changed their sponsorships within the first year, and complaints to the government were almost unheard of in the community. The workers' explanation for this was simple: they feared being deported back to India if they challenged the sponsors or revealed the truth to the government. Body-shops knew of this fear only too well and did not hesitate to exploit it. Almost all body-shops told workers that according to Australian law they had to continue with their original sponsors to remain in Australia legally. I came across only four or five workers who knew that this is untrue. But they were not sure how reliable the law was. When workers expressed discontent, some body-shops threatened to cancel the sponsorship and asked the worker to 'take the next plane back to India'. This is a typically threatening expression used on this kind of occasion by the agents. Some body-shops also intimidated workers by warning that they would write either to the immigration ministry to jeopardise the worker's future applications for PR, or to major placement agents and IT companies in order to defame the worker. A well-established firm told a worker who tried to change his sponsorship that such an action would make it impossible for the worker to join any IT professional societies in the Asia-Pacific region.

In order to make their control over workers look more legitimate, most body-shops issued 'agreements' or 'mutual understandings', which are almost completely different from the contracts that they had presented to the immigration authority on arrival. Typically, these agreements required workers to work for the body-shop for a minimum of two years. The worker had to pay a compensation amounting to three to five months' 'gross salary' (the payment that the last body-shop in the 'agent chain' receives) if he or she left before that. Some agreements required workers to give their passports and/or degree certificates to the body-shops. In some circumstances, workers were also forbidden from approaching the clients or placement agents whom they came to know through their sponsors for jobs within six months after their contracts with the sponsor were due. This was to ensure bigger agents' and clients' dependence on body-shops for labour.

But these misinformation-based measures in the information industry were not enough to control the young, educated, ambitious and even opportunistic IT professionals. A closer examination of the interaction pattern between body-shop owners and workers is needed here. Existing literature has documented some labour control measures adopted by temporary help agencies, but has focused mainly on formal policies. Gottfried (1991) identified several strategies employed by temporary help agencies to control information transmission, employee retention and management of uncertainty. In order to monitor the workers, it was found that agencies relied either on professional credentials (Cohen and Haberfeld, 1993) or steep wage progressions (Kalleberg, 2000). In the case of body-shops, however, the process of labour control was deeply embedded in a series of wider vertical and horizontal connections. The vertical connections are those between workers and body-shops through intermediaries, and the horizontal connections are those among workers. How do body-shops control workers on a regular basis through the vertical connections? And why have not workers from horizontal alliances succeeded in keeping the body-shops in check?

4.1 Vertical connections and intermediaries

Although body-shops depended on personal networks for recruiting workers, there was little of the personal touch in the everyday interactions between body-shops and workers. Both body-shops and workers emphasised to me that their relation should be 'professional', meaning that there was no need for extra interaction apart from work. Nevertheless, there was much need for constant communication, particularly when the workers were on the bench and were feeling anxious and frustrated. The communication took place through two types of intermediaries, located outside and inside the body-shops.

First, there were intermediaries who introduced the worker to the body-shop. These were either common friends of the body-shop owner and the worker, or the body-shop's business associates, or both. When there was a conflict between the worker and the body-shop, for example when the worker lost patience while waiting, these intermediaries persuaded workers to co-operate with the body-shops and even intervened to support the workers. Many workers did not expect the intermediary to help them either, and some even felt obliged by their relationship with the intermediaries not to confront the body-shop. Sam, an IT worker in his thirties, was introduced by Chitra, a friend of his based in Hyderabad, to a body-shop owned by Chaya in Sydney. Despite being on the bench for two months without pay, Sam did not blame anyone: 'I am Chitra's friend. Chitra is Chaya's friend. Chitra told me that I might have to wait for a while to find a job. He told me that the agreement [contract] might not be 100 per cent true. But if we had not made up one, I couldn't have gotten my visa in the first place. . . . I don't blame them.'

The reluctance to intervene on behalf of the workers stems from the belief of the intermediaries that it is 'unprofessional' to meddle both in a formal business transaction between the body-shops and the workers and for the sake of an informal personal relation. One informant asked: 'If I helped arrange a marriage, should I barge in when the couple fight?' The intermediaries were also constrained by the body-shop owners' position in the wider community. From the 1980s and, particularly, since the early 1990s, there was a leadership transition in the Indian community in Sydney from older physicians to younger migrants with a technical background. By the time I had completed my fieldwork, the 15 body-shops that I had observed or had close contacts with involved 30 to 40 owners and partners, of which at least 12 were active members of ethnic organisations. Many of the rest were also active participants, though they did not hold formal positions in the organisations. Body-shop owners often used ethnic functions for information exchange about their business, for example by asking other community members for good IT workers. Consequently, some community members became intermediaries. The interlocking of ethnic community leadership and body-shop ownership made it difficult for community members, whether related or not to the body-shops, to challenge the body-shops on behalf of the workers. Senior doctors and academics in the community sometimes condemned body-shops, particularly their benching practice, in private conversations with me, but there was not one single case in which established immigrants encouraged or helped a 457 visa holder to change his or her sponsorship.

More important than the intermediates from the outside, there were senior workers within the body-shop to intermediate between the owners and workers. These workers were senior in the sense that they were the first to find jobs through the body-shop or were employed by the body-shop directly.

Sameer, an employee of Advance Institute, was only 22 years old but acted in the capacity of an intermediary. He was brought in by a business partner of Kana through personal connection after the partner recognised how talented he was. Kana placed Sameer as the director of the Rubic Development Centre, the software development division of the Advance Institute, and paid him A$6000 per month after tax. Sameer told me proudly that he would intervene in the event of a conflict between Kana and the workers: 'This is good for the company. They listen. Whatever I say to the boss, the boss always says yes.'

A strong 'meritocracy' ideology prevalent among the IT workers was essential for people like Sameer to play their intermediary role. When unemployed workers complained about their situation, the body-shop owners often shut them up by pointing to the senior workers and asking: 'Can you be as good as them?' Umesh, a very dynamic and sometimes even demanding IT worker, was placed out by his body-shop and was paid A$2200 per month. Umesh estimated that the client paid his body shop A$4000–5000 for him, but he saw nothing to complain about. He pointed to a worker who was regarded as the most talented among all the workers at the body-shop. That worker was working for the body-shop as a general technology advisor for all the workers (helping the workers when they had problems at work in the client's site) with a similar salary to Umesh's. Umesh said it would look arrogant if he complained that he was being paid less. Sameer thought that the body-shopping system was fair, a view typically shared by Indian IT workers.

Another example of how the meritocracy ideology contributed to body-shops' labour control is the common arrangement that body-shops did not pay anything before a worker is placed out for the first time, but might be paid some stipend during an interim period between two assignments after he or she had worked. The logic behind this is that before the first placement the worker has not proved his or her employability and therefore the body-shop is not responsible for his or her unemployment. But after the worker has proved that he or she has the ability to be employed, the worker is treated differently from those who have not.

4.2 Horizontal connections: relations among workers

The intermediary role played by the senior workers formed only a part of co-worker relations. Compared to the talented and lucky few, there were many unemployed workers on the bench and relations among them formed a crucial dimension to body-shops' labour control, since, at least in theory, these workers could take collective actions to challenge the body-shops. However, in reality, the horizontal connections among workers tended to facilitate, rather than undermine, the body-shopping system. Assistance from fellow workers or friends helped unemployed workers survive hard times, yet did not lead to horizontal solidarity among them.

Unemployed workers often relied on fellow workers or friends to survive hard times. When I stayed with different sets of Indian IT flatmates in Sydney, ranging from four to seven, there was always at least one person who was unemployed and living off our earnings. At one time, four out of six workers were not working. Rajan, a flatmate who lived with me briefly before he found a job, took this kind of friendship for granted. Citing the situation in the US, he commented: 'There are so many young Indian IT guys in America. How can they survive there if there are no such friends? [. . .] Western people have no sense of community. Their "mates" are not real "mates".' Some informants told me that local workers in the US are more reluctant to move to the San Francisco Bay area, a highly expensive region, than Indians. This is because Indians can cut their living costs by sharing flats but Americans prefer to live alone. One informant explained to me why Indians in the US usually get married as soon as they get H-1B visas. If they were to remain single they would face endless requests for financial help from friends: 'If you are married, your friends can't ask for money from you.'

However, the support networks among the workers did not win workers more bargaining power against body-shops. A typical example of the lack of solidarity and collective actions among workers is how Uday and Sree Kumar, IT workers from India and very close friends, handled pressure from their body-shop. Uday was introduced to another body-shop by Sree Kumar, when they were both on the bench, and subsequently both of them got jobs through an agent. Kana, the owner of the body-shop where Uday worked, soon became aware of this and asked Uday for 'compensation'. Uday agreed to pay Kana A$1000 per month for as long as he continued working. Uday believed that Kana must have asked for compensation from Sree Kumar as well, but, surprisingly, Sree Kumar and Uday never talked about this at all.

The highly flexible labour market entailed a very uncertain and unclear working situation. Workers did not know when they could have a job, when they might be unemployed, how much they would be paid and how much they should ask for. Negotiations between workers and body-shops were always carried out on a one-to-one basis. Without general standards, it was difficult for the workers to discuss their situations, let alone to take collective actions. Furthermore, there was a widely held view that to ask about other workers' pay and attempt to ascertain how salary was decided on was 'unprofessional'. In this sense, professionalism reduced transparency, rather than enhancing it.

The lack of transparency is further compounded by a peculiar tendency of secrecy among young Indian IT workers. The workers were so wary of attracting fellow workers' jealousy[7] that they kept their intentions as well as their problems closely guarded secrets. When Uday applied for his PR, he did not tell this to his cousin who was in Sydney. He explained: 'If I tell this to someone, I will fear something may happen. We feel that the person may write letters to DIMA [the Australian immigration ministry] saying I have a bad character.' Uday's brother, also one of our unemployed flatmates, found a job with Lucent Australia and Lucent Australia sent the brother to the US for training. But the flatmates were told that he was going to the US to look for jobs. Uday said: 'We can't trust these people. If they know the truth, they may pray to God to destroy our prosperity.' Workers' sensitivities were so strong that I had difficulties adopting a snowballing method to find new interviewees. People were reluctant to introduce other workers to me. More than once, informants called me after interviews to seek my assurance that the information they shared with me would remain absolutely confidential.

The tendency of secrecy and the sensitivity surrounding money seem to be 'cultural' issues, but in a certain institutional context they served to enhance the contradictions of the IT industry. The uncertain working condition on one hand created secrets for each worker, and on the other hand generated an atmosphere of fear. Although IT professionals of other origins may feel more comfortable in talking about money in general, they rarely know about colleagues' pay and can rarely take any collective actions to acquire better wages and living conditions from their employers.

5. Worker strategies: 'quiet exit', 'open lift' and résumé forwarders

Workers did not always conform to the established structure, with some attempting alternative means to secure placement by taking enormous risks. In order to have a full picture of the labour control mechanism in the global IT industry, we must examine workers' strategies. Apart from waiting for jobs from the body-shop, workers on the bench also sought opportunities by strategies of what I call 'quiet exit' and 'open lift'.[8] Quiet exit means the worker's searching for opportunities outside the body-shop without challenging or even informing their sponsors, while open lift means leaving a body-shop openly after obtaining a PR in Australia or after getting a visa to other countries, particularly the US, around the time the worker's 'agreement' was due to expire.

What Uday and Sree Kumar did, that is, getting jobs through other agents behind their sponsor, is a typical 'quiet exit' strategy. They did not change their sponsorships, but the second agent projected himself as the sponsor of Uday and Sree Kumar in order to get their wages from the client and the bigger agent. This was a risky game. Everyone involved would be in trouble if it was found out by the client or the authorities. Uday, Sree Kumar and the second agent took this course of action because they did not want to deal with Kana with regard to the change. Uday had another reason to lie and to be 'quiet'. He believed that the change of sponsorship would handicap his future application for PR.

Venkate, a former salesman in one of India's largest insurance companies, was one of the most sophisticated informants I came across. When he was on the bench at a body-shop, he read newspapers everyday and visited immigration solicitors, finance consultants and lawyers to see how he could become a PR as soon as possible. He got his PR nine months after arriving in Australia. The strategy of quiet exit opened up a way for workers in difficulties, but it did not challenge the body-shop system in any sense.

Only a few workers could make a quiet exit, so the majority hoped for an 'open lift'. When a worker left a body-shop either because he or she obtained an Australian PR and therefore did not need sponsorship any more, or because he or she got an opportunity in another country, particularly the US, the body-shop was normally quite happy to let the worker go even if that happened before the contract was due to expire. In some cases, small farewell parties were organised by the body-shop. The body-shop did this partly to show to other workers, particularly those on the bench, that everyone would have a bright future after the agreement.

Unlike quiet exit, which led to tension between the body-shops and the worker, open lift did not necessarily mean that the worker would cease their connections with the body-shop. Some workers, in fact, became closer to their former sponsors, mainly by acting as 'résumé forwarders'. Résumé forwarders were middlemen between body-shops and potential workers. They helped workers who wanted to migrate to Australia or to change jobs or sponsorships in Australia by forwarding their résumés to body-shops. All body-shops were connected to résumé forwarders, though the number of the résumé forwarders is difficult to estimate since most workers acted occasionally in this capacity.

The majority of the résumé forwarders were former 457 visa holders who became recent PRs, and they understood this role as part of their career advancement or a step forward towards setting up their own business. Young Indian IT workers were in general very entrepreneurial, very unlike earlier skilled migrants who preferred secure jobs. Stories of successful Indian IT businesspersons from both the US and India were widely circulated among them: 'If they can do it, why can't I?' Another informant told me that he has 'something burning in the heart, a sort of anger' to run a business of his own. When moving up the career ladder from technicians to entrepreneurs, setting up body-shops is the most effective means for many Indian IT workers to accumulate their initial capital, and acting as résumé forwarders was seen as a good warming-up exercise for this.

The immediate return for acting as a résumé forwarder was money in the form of commission from body-shop owners. Résumé forwarding also helped to build up networks and accumulate experiences needed for future business. Résumé forwarders not only passed on workers' résumés, but also forwarded information about new openings, often obtained from the company where they were working or big agents who employed them, to the body-shops. The body-shops also asked forwarders to search job opportunities or any short-term project for their workers on the bench. Indian body-shops normally kept at a distance from each other because of the sensitivity stemming from their undertaking the same business, and sometimes even asked résumé forwarders to collect information about each other. Since the IT placement business was based on 'agent chains', finding a place in the chain was critical for setting up one's body-shop.

The résumé forwarders, in turn, assisted body-shops in the control of workers. When body-shops paid forwarders commission for introducing workers, the commission was paid only six months after the worker was placed out. If the worker could not be placed out or changed his or her sponsorship within the six months, the forwarder would receive nothing. Therefore, it was certainly not in the forwarders' own interest to see workers jump ship. When body-shops 'sold' sponsorships to workers, particularly those without proper IT skills, through forwarders, the forwarder was supposed to prevent the worker from turning the tables to sue the body-shop for not offering him or her the job as indicated in the ostensible contract.

Similarly to the internal intermediaries in body-shops, résumé forwarders' control of workers was, to a large extent, based on workers' respect for them. Generally, the workers were supposed to be grateful to the forwarders for giving them new job opportunities. It was common for workers to send sweets to the forwarders and invite the forwarders to the *puja* that they sponsored in Hindu temples.[9] Yet, unlike the intermediaries, résumé forwarders had a clear goal in sight: to set up business. Therefore, they were keener to maintain good relationships with the body-shops and, consequently, were stricter with workers.

6. Discussion and conclusion

Labour control mechanisms in the global IT industry can be seen at macro, micro-institutional (internal relationships in body-shops) and individual levels. Workers' strategies for moving up the career ladder clearly demonstrate that these three mechanisms form a full circle to make the system self-perpetuating. As to the question of how labour is managed and controlled in a highly uncertain situation, the answer may appear paradoxical: labour is controlled by manipulating, often by exacerbating rather than reducing, uncertainty. Central to the labour control mechanisms is a process to transfer uncertainty, or, more accurately, the labour cost borne of uncertainty, from one party to another. IT companies transfer the labour uncertainty to the placement industry, large placement agents transfer the uncertainty to small agents, and small agents ultimately transfer the cost to the workers. Workers were willing to abide by the system because they believed that uncertainly brought with it new opportunities – without uncertainty and the consequent demand for a large labour pool in developed countries, Indian IT workers would have fewer chances to chase their IT dreams abroad in the first place. Uncertainty was believed to be accompanied by more opportunities for unconventional career development (for example, being promoted to CEO in one's twenties) and for small business startups (by which one may become a millionaire overnight). By exploiting such uncertainty, workers, in turn, help body-shops to control new workers, and some further turned themselves into fully fledged body-shops, and, therefore, the agents of labour control.

No less important than the institutional arrangement of labour control is workers' perception of uncertainty. Uncertainty on the one hand creates fears, leading to sensitivity and jealousy that subsequently undermine solidarity between workers, yet on the other hand it encourages meritocracy and 'professionalism', since these ideologies make the uncertain world more comprehensible. Uncertainty also induces enterprise and entrepreneurship as common aspirations among many IT workers.

All these notions are at the centre of the hegemonic discourse in the era of globalisation. Knowledge workers are deeply convinced that uncertainty is and should be part of our life, and their career goal is to achieve something unimaginable in traditional industries. Traditionally, the working class was said to be unable to understand class relations beyond their immediate surroundings. That is, workers identify individual capitalists as their problem, rather than capitalistic institutions. Indian IT workers justify their concrete situations by accepting abstract rules, particularly the rule of uncertainty, which were thought to be beyond anyone's control – be it the workers, the body-shop owners, or even the IT majors. The capital–labour divide was believed to be no longer valid. IT professionals referred to themselves as 'consultants', which may mean body-shop owners, workers managed by body-shop or freelance specialists, instead of 'workers' or 'labourers'. Words such as 'lay-off' or 'fire' were rarely heard in their daily discourse. The 'correct' phrases for firing were 'rationalising', 'right-sizing' and 're-engineering'. This hegemonic discourse not only made workers docile but also concealed the issue of labour relation in the IT industry altogether.

Finally, it must be stressed that what this chapter has explored, that is, how the system sustains itself through processes internal to it, is only part of the story. Uncertainty and labour control mechanisms created a few winners but many losers. Someone outside of that system must shoulder the cost in order to sustain that system. The boundary between the IT industry and other sectors may be more important than that between labour and capital within the IT sector. In other words, the IT industry became a game where those able to get in, with either capital or knowledge, may become big winners, yet the question remains, from where does the initial financial capital and human resource originate and how are the losers' costs absorbed? This necessitates a study of the IT industry in a broader context than that attempted in this volume. The particular issue regarding the production of the IT labour force within the overall political economy of information capitalism dealt with in this chapter is intended to add to the overall debate.

Notes

1. During the IT market slowdown in 2001, 'B2C' was jokingly referred to as 'Back to Chennai' and 'B2B' as 'Back to Bangalore'.

2. Java is a software program created by Sun Microsystems to allow www pages to contain a code that is executed on the browser.
3. Interviews with Terry Porter, ICON Placement Ltd, 11 May 2000, Sydney; Stephen Leo, MBT Ltd, 11 May 2000, Sydney; Michael Bulter, APESM, 6 March 2000, Melbourne.
4. Interview with Michael Bulter of ITPA, 6 March 2000, Melbourne.
5. This agency is now part of the US Department of Homeland Security.
6. This cost includes the visa application fee, airfare and, in some cases, short-term accommodation. The fee paid by workers before leaving India is taken by the Indian body-shops.
7. Jealousy or envy is a very important concept in the everyday life of many parts of Indian society. In Telugu, *asuya* and *disty* mean envies with destructive effects. The wariness of others' envy makes people secretive. In Andhra Pradesh, when people achieve something that cannot be hidden, such as a new house, a car or other status symbols, special *pujas* are conducted to avoid others' *asuya* or *disty*. Srinivas (1996, pp. 183–6) details how envy is perceived and how people avert it with great effort in Karnataka.
8. I coined these two terms by following Hirschman (1970), who summarised two basic strategies of how people address dissatisfactions with an organisation in a market economy: 'exit' and 'voice'. 'Exit' refers to the strategy that employees or customers adopt to leave an organisation, which is essentially 'voting by their feet'. 'Voice' implies members raising their concerns in the organisation without leaving, which is, essentially, a political action. As mentioned earlier, 'voice' was effectively discouraged in body-shops and quiet exit and open lift can be seen as sub-categories of 'exit' as proposed by Hirschman.
9. It is an Indian custom to hand out sweets during celebrations, particularly to show gratitude and respect to those persons who helped achieve the success. People sponsor *puja*s in temples for special family events. The *puja*s are open to everyone, but the family may invite a few special guests.

10
The Political Economy of Informational Development: A Normative Appraisal

Govindan Parayil

1. Introduction

The rapid advances in information and communications technologies (ICTs) during the past three decades, acting in tandem with the globalisation of trade and investment, have brought about unprecedented changes in the global economy. The discovery of 'knowledge'[1] as the most important ingredient of all modern economies, surpassing land, labour and capital, has profoundly affected the modes of global economic production as well as consumption. Unleashing an array of new products and services for which usability was mostly invented, and reviving the productivity potential inherent in several advanced sectors of the economy by the deployment and diffusion of ICTs, represents the key factor behind this transformation. In advanced countries where ICTs have become part of the economic infrastructure, firms and business organisations, regardless of their size or spatial–temporal location, have experienced steep decline in transportation, communication, search and distribution costs (Guha, 2003). That is, the development and diffusion of ICTs have become a source of comparative advantage of nations (James, 2001). The assumption, until now, that the standard Heckscher–Ohlin model of international trade where comparative advantages in factor and resource endowments held sway has been replaced by the comparative advantage driven by international differences in production functions due to innovation and rapid technological change (Krugman, 1995). The advances in ICTs and related technologies have given rise to a new form of comparative advantage and international division of labour in the global economy. These economic changes may be collectively identified as 'informational development' because of the centrality of knowledge as codified information through the mediation of ICTs. Informational development is a complex and uneven process of economic change due to the dialectical relationship between the global–local forces and factors that underpin its shaping. It has spawned a host of normative and equity questions.

This final chapter reflects on these questions by probing the contradictions inherent in informational development. However, the basic approach of the chapter is generally to reflect on the normative implications of informational development within the contours of knowledge capitalism, which is viewed within the context of the gradual shifting of the global economy from industrial capitalism to information capitalism (Bell, 1973; Touraine, 1969; Drucker, 1969,1993; Stehr, 1994; Schiller, 1999; Castells, 1996; Burton-Jones, 1999).

To understand the unravelling of industrial capitalism and the rise of informational capitalism, one needs to look at the way in which the meaning of knowledge itself underwent change as well as the changing relationship between capital, labour, machines and knowledge. With the widespread adoption of automation in production since the 1970s, scholars such as Ernest Mandel (1975, 1978) predicted the 'inner limit of capitalism', a long-held Marxian position. However, that capitalism proved itself a dynamic socio-technological system that had overcome many other previous challenges to its survival showed that it can not only coexist with the advent of automation and fast-changing information and communication technologies, but also, in fact, thrive under this new productive relationship. Morris-Suzuki (1997) has shown that advanced industrialised countries have co-opted automation and 'softened' their economic conditions to extract more output and economic value from non-material elements such as research, science and innovation. What this transformation shows is that knowledge can be separated from traditional factors of production such as labour and physical capital. As Arrow (1962) observed long ago, invention and innovation are basic acts of knowledge production. Knowledge turned into intellectual capital can exist as an independent commodity and in turn become the most important factor of production from which more value can be extracted under capitalist social relations of production and exchange. The ascendance of knowledge as the pre-eminent form of capital along with the rise of a new synthetic labour category called 'knowledge worker' (who is part biological labourer and part source and repository of intellectual labour) created new economic relations that thwarted the social contract built up between labour and capital in the industrial age. The unravelling of social democracy and the welfare state due to the rise of neo-liberalism saw the slow death of socialised labour.

The 'knowledge workers', though still exploited for surplus intellectual capital, became free agents and consultants. Labour as a social category of industrial capitalism took on a new meaning in informational capitalism, where wage-labour began to look like a relic of the past. Labour-time as a social and place- and time-bound entity became a spatially and temporally independent commodity that can be place- and time-shifted through the mediation of ICTs. This is the key factor behind off-shoring or outsourcing of knowledge (service) work. Knowledge as an abstract public good turned into an abstract intellectual property and the ultimate source of wealth. Ultimately, it was the reification of knowledge as well as the increased commodification of knowledge and mental labour as intellectual property that provided the key to information capitalism.

We can see two contradictions of ICT-led development – increasing returns and digital divide. Therefore, the main argument of this chapter is that the digital divide, income dispersion, uneven development and asymmetries in intra-regional economic growth in parts of the developing world and the deepening income inequality in the developed world after the onset of the global information revolution are not unrelated or mutually exclusive phenomena. They form an integral part of the economic restructuring occurring after the onset of economic globalisation. Therefore, the so-called digital divide, which has morphed in the context of development discourse to become the most frequently cited problem concerning uneven development, is not an accessibility issue as it is being made out to be, but, essentially, an equity issue.[2] Most analysts look at the problem of digital divide in a narrowly construed instrumental sense of the absence of a technical artifact for a large number of people in the world that ostensibly deprives them from being a part of the so-called 'knowledge economy'. The digital divide is often portrayed in crassly reductive terms as a mere technological access problem that can be ostensibly addressed by providing cheap computing and communication technologies to the poor. However, the digital divide is not merely a technological problem due to the absence of connectivity or lack of access to cyberspace. This instrumentally informed discourse on digital divide is a modernist tendency to unreflexively categorise and compartmentalise complex socio-technological changes into one-dimensional social problems in a bid to resolve them through simple technological fixes (Parayil, 2005).[3]

There is an asymmetrical relation between traditional means and methods of production (dominated by primary, bulk-processing and manufacturing industries) and the innovation- and knowledge-driven industries of 'post-industrial weightless economy'. While the production functions of the former are characterised by constant or diminishing returns, the latter enjoys the benefits of increasing returns. The distributional inequities of these dual modes of production are manifested as the digital divide and the extreme dispersion of income between the so-called knowledge workers and unskilled workers the world over (Parayil, 2005).

2. Informational development and the new international division of labour

Flexible production, free movement of capital on a global scale in real time and the proliferation and processing of huge amounts of information as well as the commodification of knowledge are the characteristic features of the 'knowledge-based economy'.[4] The replacement of industrial capital by financial capital; its free flow through global information networks in real time without being impeded by spatial and temporal barriers; and the intensification of trade in goods and services on a global scale are its other key features. Whereas the previous industrial economy was made up of 'space of places', the new global economy is characterised by the 'space of flows' of free-floating capital, consumer goods, cultural artifacts, fashion, music, highly skilled people, refugees, criminal enterprises, drugs and weapons in a seemingly 'borderless' but networked world (Castells, 1996). However, places are still important, as Dirlik (1999) has shown, because places are important sites of resistance, a process accentuated by the dialectical relationship between the global and the local axes of globalisation. As Henderson et al. (2002, pp. 437–8) show, 'the world is now constituted by *both* a space of places *and* a space of flows' (emphasis in original) and 'places are being transformed by the flows of capital, labour, knowledge, power, etc'. But, as Castells (1996) has argued, it is precisely the networked and global spaces of flows spearheaded by information technologies that shape knowledge capitalism.

As has already been shown by numerous scholars, there has been a progressive dematerialisation of the economy, from the commodities of industrialism with high congealed material resources to the new high-tech commodities of informational capitalism whose value chain is determined by the informatisation of production and exchange (Pohjola, 2001; Quah, 2001). It has also become evident that while agriculture, extraction and bulk-processing part of the global economy are still caught up in the iron law of diminishing returns, high-technology and knowledge-based industries enjoy increasing returns and disproportionately higher value added to their economic output (Arthur, 1994). While the prices of agricultural and mass-produced commodities and raw materials have plummeted worldwide, the relative value of information-based commodities has been rising steadily in relative terms of trade.[5] Therefore, it is important to ask why the value chain of traditional economy is different from the value chain of the new informational economy. Why does economic globalisation redefine the value chain of primary and industrial production based on their spatial location but fail to do the same for the informational economy? Why is it that a farmer's or a manufacturing worker's productive work is not treated as knowledge work while those connected to the network of informational capitalism are treated as knowledge workers? Unlike the iteration of the capitalist mode of production punctuated by previous instances of technological clusters and associated technological changes, the introduction of ICTs and other modern technological innovations has caused a profound rupture. It is precisely this rupture that makes the digital divide and other distributional inequities of informational capitalism such important issues to be analysed within the context of the political economy of globalisation, rather than be seen in isolation as a mere technological access problem.

This rupture has several ramifications. The most important is the inauguration of a new international division of labour that began in the 1970s, ostensibly the time that marks the beginning of the 'Information Age' (cf. Castells, 1993). As advanced industrialised nations began their transition to their post-industrial phase, there was a progressive shift of manufacturing industries to the developing world. During the two decades between 1970 and 1990, more than 60 per cent of manufactured goods came from developing countries (Cohen, 1998), and this percentage has increased dramatically since China became the most important manufacturing centre of the world, especially since the 1990s. The shifting of manufacturing to the developing world caused profound concern in the developed world of possible de-industrialisation. But as Cohen (1998) argues, unlike the de-industrialisation of the colonies due to the importation of manufactured goods from the imperial powers, the importation of manufactured goods from the developing world to the developed world did not cause the same kind of de-industrialisation that the colonies had suffered earlier (such as in colonial India). De-industrialisation of heavy industries such as steel, ship-building and mining did take place on a large scale in countries such as the United States and Britain.[6] While de-industrialisation became the reality for sunset industries, we saw an entirely new form of industrial transformation involving information, communications, computers, material and biotechnologies taking place in the developed world. This 'tertialisation' of the economy saw a profound structural transformation of the labour and factor markets and its consequent impact on income distribution.[7]

While low value-added production moved to cheaper production destinations in the South (to take advantage of its cheap labour, land, and certain capital resources)[8], the North held on to the patents, innovations and production techniques of technologies from which it derived more value than by producing the goods themselves.[9] While the manufacturing industry shrank in the North and consequently the number of jobs and the income of the working class in that region decreased, the income of the so-called knowledge workers involved in the new informational economy increased by many times.[10] It was no accident that the coming of the 'weightless economy' saw extreme dispersion and distribution of income and wealth along with the most spectacular increase in global wealth (specifically concentrated among certain individuals, regions, countries and corporations).

The income and wage distribution in the US and other OECD countries since the 1970s (the beginning of the age of information capitalism) changed unexpectedly, contrary to the trends since the post-war period. In these countries income inequality has risen sharply and there was a sharp decline in the demand for less-skilled workers and a consequent high demand and wage increase for highly skilled workers. This phenomenon is an important aspect of informational development because inequality had been declining in most industrialised countries up to the 1970s. It has been noticed that all economically advanced states started out the process of industrial development with low income inequality and ended up with progressively worsening inequality as industrialisation and economic growth picked up steam. As economic growth continued and as more and more workers entered the high productivity sectors of the economy, income inequality began to recede with increasing income (Kuznets, 1955). Kuznets's prediction of this reverse causal relationship between growth and inequality had been validated for the US and most of the OECD countries up to the 1970s (Aghion et al., 1998). Ahluwalia (1976) found the same reverse causal relationship between growth and income distribution among developing countries.[11] A corroboration of Kuznets's hypothesis in the case of the US could be adduced from the historical fact that the share of total wealth owned by its top 10 per cent richest households rose from 50 per cent around 1770 to 70–80 per cent around 1870, and then receded back to 50 per cent around 1970 (Aghion et al., 1998, p. 9). The sharp reversal of income inequality in the US and other industrialised countries since the 1970s is appropriately portrayed as the Kuznets U-curve doubling back on itself (Picketty and Saez, 2003). The top 1 per cent wealthy people's share of income increased from 5 per cent in 1970 to 11 per cent in 1998 in the US (Picketty and Saez, 2003). Along with the rise in income inequality, the US also increased its share of global wealth compared to other nations and regions in historical terms, as shown in Tables 10.1 and 10.2.

It is an astonishing historical footnote (with little material significance to the current global economic structure) to note that around 1700, India had 25 per cent of world GDP, while the (would be) US had

Table 10.1 Share of world GDP of selected countries and regions, 1500–2001 (percentage of world total; GDP level of each unit normalised on 1990 international dollar)

Year	USA	UK	Western Europe	Japan	India	China	Mexico	Africa
1500	0.3	1.1	17.9	3.1	24.5	25.0	1.3	7.4
1700	0.1	2.9	22.5	4.1	24.4	22.3	0.7	6.6
1820	1.8	5.2	23.6	3.0	16.0	32.9	0.7	4.5
1870	8.9	9.1	23.6	2.3	12.2	17.2	0.6	3.6
1913	19.1	8.3	33.6	2.3	7.6	8.9	1.0	2.7
1950	27.3	6.5	26.3	3.0	4.2	4.5	1.3	3.6
1973	22.0	4.2	25.7	7.7	3.1	4.6	1.7	3.3
1998	21.9	3.3	20.6	7.7	5.0	11.5	1.9	3.1
2001	21.4	3.2	20.3	7.1	5.4	12.3	1.9	3.3

Source: Maddison, 2001 and 2003

Table 10.2 2001 share of world income of selected countries and regions, real and purchasing power parity (PPP) adjusted

Type	USA	UK	South Asia	Japan	India	China	Mexico	Africa*
Real	31.4	4.6	2.0	14.5	1.5	3.6	1.7	1.0
PPP	21.3	3.2	6.8	7.5	5.5	11.7	1.9	2.4

Source: World Bank, 2003b
Notes: *For Sub-Saharan Africa

a global GDP share of just 0.1 per cent and the UK had only 2.9 per cent. Three hundred years later the role has reversed, with the US commanding 31.4 per cent of global income while India's share shrank to a miniscule 1.5 per cent (Table 10.1). See also Tables 10.3, 10.4 and 10.5 for global trends in income per capita in historical terms.

A secular rise in income inequality is visible since the 1970s not only in the US and OECD countries, but also in developing countries like India, Mexico, China and Brazil as shown in Table 10.6. It is particularly noticeable that inequality was declining in these countries from

Table 10.3 Ratio of GDP per capita of ten highest to ten lowest countries: 1950–2001

Year	1950	1960	1970	1980	1990	2000	2001
Ratio	36.2	33.9	32.7	32.2	34.2	47.0	47.2

Source: Sutcliffe, 2004, based on Maddison, 2003

Table 10.4 GDP per capita of selected countries and regions: 1500–2001 (1990 international dollar)

Year	USA	UK	Japan	India	China	Mexico	Africa	World
1500	400	714	500	550	600	425	400	565
1700	527	1,250	570	550	600	568	400	615
1870	2,445	3,191	737	533	530	674	444	867
1913	5,301	4,921	1,387	673	552	1732	585	1,510
1950	9,561	6,907	1,926	619	439	2365	852	2,114
1973	16,689	12,022	11,439	853	839	4,845	1,365	4,104
1990	23,214	16,410	18,789	1,309	1,858	6,097	1,385	5,154
1998	27,331	18,714	20,418	1,746	3,117	6,655	1,368	5,709
2001	27,948	20,127	20,683	1,957	3,583	7,089	1,489	6,049

Source: Maddison, 2001 and 2003

Table 10.5 2001 per capita incomes of selected countries and regions, real and purchasing power parity (PPP) adjusted

Type	USA	UK	Japan	India	China	Mexico	Africa*	World
Real	34,870	24,230	35,990	460	890	5,540	470	5,140
PPP	34,870	24,460	27,430	2,450	4,260	8,770	1,620	7,570

Source: World Bank, 2003b
Notes: *For Sub-Saharan Africa

Table 10.6 Gini Index (in percentage) of selected countries: 1950 and after for various years

Brazil										
Year	**1960**	**1970**	**1976**	**1980**	**1985**	**1989**	**1992**	**1995**	**1996**	**1998**
Gini	53.0	57.6	60.0	57.8	58.9	57.0	58.1	59.9	60.1	60.7
USA										
Year	**1950**	**1960**	**1965**	**1970**	**1975**	**1980**	**1985**	**1990**	**1994**	**1997**
Gini	37.9	36.4	35.6	35.3	35.7	36.5	38.9	39.6	42.6	45.9
China										
Year	**1953**	**1964**	**1970**	**1975**	**1980**	**1985**	**1987**	**1990**	**1992**	**1995**
Gini	55.8	30.5	27.9	26.6	32.0	31.4	34.3	34.6	37.8	45.2
Mexico										
Year	**1950**	**1957**	**1963**	**1968**	**1977**	**1984**	**1989**	**1992**	**1994**	**1996**
Gini	46.0	50.0	49.0	49.0	50.0	50.7	53.1	53.4	56.7	58.0
India										
Year	**1950**	**1960**	**1965**	**1968**	**1975**	**1983**	**1986**	**1990**	**1994**	**1997**
Gini	41.0	47.3	42.8	46.3	40.5	31.5	30.5	29.7	34.5	37.8

Source: WIDER, 2000 and World Bank, 2003b

the 1950s till the 1970s.[12] But this secular trend in the decline of inequality began to reverse course from the late 1970s in industrialised countries and from the 1990s in developing countries like India, China, Mexico and Brazil. The information revolution that began in the mid-1970s is attributed both to this trend and to the doubling of the share of wealth of the top 1 per cent since 1976, in addition to the overall economy-wide rise in income inequality in the US (Wolff, 2002). The Information Revolution, however, gained momentum from the 1990s with advances in computers, communications and biological technologies, further affecting the pace of income and wealth creation and their distribution in the global economy.

The rapid pace of computerisation of the workplace; the diffusion of ICTs in the production processes and organisational settings; and the concomitant high premium paid for cognitive skills of those with higher education are the prime factors responsible for the sharp rise in the wages and earnings of the well off (Bresnahan, 1997; Autor et al., 1998; Wolff, 2002). Rising wage inequality is attributed to rapid technological change in the ICTs cluster, specifically due to the large-scale diffusion and deployment of computers and IT equipment in the US economy. Rapid decline in the price of computers and an ensuing surge in computer diffusion had a pronounced effect on wage dispersion due to changes in the production function (Krueger, 1993). Stiroh (2002) shows that aggregate labour productivity accelerated in industries that either produced IT or used IT most intensively in their production function. Autor et al. (1998) attribute two distinct features associated with technological change to the wage dispersion. First, the long-run secular increases in the relative demand for highly-skilled workers are associated with: (1) biased technological change and the consequent changes in the production processes; (2) large-scale organisational changes at the workplace and shop floor level; and (3) capital deepening. Second, the impact of the technological change was felt in the strong demand for more skilled workers and the decline in demand and wage for unskilled workers. Evidence of pervasive skill-biased technological change was also found internationally, especially in OECD countries (Berman et al., 1998). One of the significant changes that occurred during this process of biased technological change was the large employment shift towards non-production workers (Davis et al., 1991; Berman et al., 1994). Wolff (2002) shows that the ratio of GDP to PEP (persons engaged in production) declined from 2.58 during the 1947–76 period to 0.93 during 1979–97 in the US, a drop of nearly three times.

Declining unionisation in both industrialised and industrialising countries contributed to the lowering of wage shares of low-skilled workers because unions with low membership lost their power for collective bargaining on behalf of the workers (Gordon, 1996). In 1997, only 14.1 per cent of workers in the US were members of labour unions (Wolff, 2002). Reich (1991) argues that 'symbolic analysts' who use IT equipment and produce knowledge have been able to increase their incomes due to the advantages accrued to them consequent to economic globalisation. Wood (1995) and Green et al. (2001) attribute the globalisation and liberalisation of trade as an important contributor to the declining employment and income levels of low-skilled workers during the 1980s and 1990s. After joining NAFTA and opening its economy to technology, trade and investment, Mexico's trade volume with its northern neighbours multiplied by many times. However, only some 10 per cent of Mexicans at the top of the income pyramid gained most of the new wealth after globalisation and liberalisation of Mexico's economy (Thompson, 2002). Income and wealth inequality became so acute that the Mexican middle class has pretty much disappeared. This is in a country where middle and working classes made up some 60 per cent of the population in the 1970s.

Lucas and Sylla (2003) argue that the spectacular innovation clusters centred on the Internet and other information technologies show signs of 'further increasing global inequalities' in an already unequal world. Antonelli (2003) argues that the digital revolution is 'engendering a global digital divide because of composition effects'. That is, nations and regions that are equipped with the infrastructure to adopt ICTs in their production functions and organisational settings are further advantaged by this ability in terms of sharp increases in their productivity levels and international competitiveness. Those nations and regions that cannot climb onto the informational bandwagon are further disadvantaged because of their inability to compete, a clear case of increasing returns and positive feedback phenomenon at work in the economy. This compositional effect is not confined to developed nations alone, but also to regions within some developing nations. The sharp rises in their productivity levels associated with the ICT revolution caused severe income dispersion within such regions, and the international competitiveness of technologically advanced states allowed them to increase their share of global wealth *vis-à-vis* the poor countries or regions that were left out of the ICT revolution. ICTs have become the most potent instrument for integration into the global economy, and the poor integration of these technologies into the over all economy is another key element in the cascading cumulative causation of events leading to further marginalisation of the marginalised (James, 2001).

3. The digitisation of inequality

While great wealth was created for those connected to the winning camp of globalisation and the knowledge-based economy, it also spawned the most unequal distribution of income and wealth in human history. The average income differential between peasants and manual workers and those we band under the category of 'mental' workers (estate managers, teachers, physicians and priests) in feudal societies was about 1:100.[13] It was an extremely unequal society with a weak central state that vested the real power in the landed feudal lords and the aristocracy. When these feudal societies were transformed by capitalism into industrial societies, the income inequality between the industrial workers involved in manufacturing and those involved in 'mental' work (the professional class made up of managers, physicians, engineers, designers and university teachers) declined considerably. Williamson and other prominent economic historians have established the phenomenon of a general convergence and narrowing of income disparity since the beginning of the Industrial Revolution till the 1970s.[14] Despite the iniquities of the hard times faced by the working class of industrial societies,[15] the Industrial Revolution had a levelling effect on socio-economic and political differences (between classes) compared to its much more iniquitous agrarian/feudal predecessor. The income differential between the industrial workers and the elite groups has reduced to about 1:10, relative to the feudal epoch (Parayil, 2005).[16]

As industrial societies began to enter the post-industrial stage with the onset of the so-called knowledge-based economy that is being constantly challenged by the creative destructive forces of Schumpeterian innovators, the income differential between those who do physical ('menial') work and those who do mental ('knowledge') work became worse than that of the feudal period. We can reasonably argue that the income differential between unskilled workers and the knowledge workers and 'symbolic analysts' has widened to an astonishing 0:1 ratio.[17] This is precisely what is meant by the claim that you are either 'in or out' of the 'winner-take-all' knowledge economy. The significance of Drahos (1995) and Drahos and Braithwaite (2002) citing similarities between feudalism and informationalism makes sense in this context – the rent-seeking behaviour of knowledge economy firms (by ratcheting up the intellectual property right claims) is essentially the same as the rent-seeking feudalists. Although the real amount of relative wage earned by the unskilled workforce is not exactly zero, one can imagine the income differential between those in the knowledge economy and those out of it to be extremely high. What is being argued is that the knowledge elite takes a disproportionate share of the wealth of the digital economy in the information age. This is a clear example of the income inequality bred by the 'winner-take-all' markets of globalisation.[18]

As pointed out in Table 10.7, the income differential in the knowledge (digital) economy, ironically, follows the (digital) logic of Boolean algebra. This is the prospect that the digital divide and the new international division of labour have for those people and nations that are out of the new economy. While Marx and Engels spoke, during the 1850s, of the technologies and markets of industrial capitalism knocking down the Chinese Walls of 'barbarian' and feudal societies, the information, communications and biotechnologies of the information

Table 10.7 Notional income and wage for manual work and knowledge work in historical perspective

Mode of production	Nature of the economy	Wage for manual/ physical work	Wage for knowledge/ symbolic work
Agrarian/feudal economy	Primary/ extractive	1	100
Industrial economy	Secondary/ manufacture	1	10
Informational economy	Tertiary/service/ symbol analysis	0	1

Source: Parayil, 2005

age are compelling all nations, under the threat of irrelevance, to join the bandwagon of economic globalisation.

4. Increasing returns and the dynamics of the digital economy

Economists from the classical period to modern times had struggled to explain how productivity growth (key to income and wealth generation) had actually happened. Smith, Ricardo, Marx and others provided the classical explanation of economic growth and wealth accumulation. The general consensus is that pushing the production possibility frontier through rapid technological change is the means by which nations generally develop (Parayil, 1999). Mokyr (1990, pp. 4–6) summarises four factors responsible for this process of economic change and growth:

1 Capital accumulation: when capital stock accumulates faster than the growth of labour force, the capital per unit of labour increases. Therefore, the average worker has more tools and equipment at his disposal and hence his productivity increases.
2 Commercial expansion: increase in the exchange of goods, services, labour or capital can be beneficial to all players involved.[19]
3 Scale or size effects: population increase and a large population base are conducive to economic growth, contrary to the popular neo-Malthusian view that large populations are inimical to economic growth.
4 Increase in the stock of human knowledge: advances in science and technology, institutional innovations and overall technological progress help to improve productivity.

Ultimately, the individual or combined effect of these four factors is reflected in the rate of technological change that a nation undergoes, which ultimately determines its level of prosperity. However, entrenched political economic policies and structural conditions of the economy, largely, determined the distribution of income and wealth and access to education, skills and technology.

Long before neoclassical economists 'explained' the hidden logic behind productivity growth and wealth creation, Joseph Schumpeter had challenged their claim that economic change and growth occur only when existing factors of production are optimised. The idea that any intrusion of external factors in the economy should be avoided was shattered when Schumpeter argued that it was the innovating entrepreneurs who disrupted the economic equilibrium through technological innovation that actually brought about economic change.

But what constitutes innovation, and how to create the environment conducive to innovate? It is more or less agreed that a society's or firm's ability to create knowledge is crucial to technological innovation. Technology is essentially society's repository of knowledge regarding industry, agriculture and other productive ventures (Mansfield, 1985). That is, technology is essentially knowledge (Layton, 1974). Knowledge is a cumulative resource that does not diminish when others use it because it is a non-rivalrous and non-exclusive entity. In fact, its value and content are enhanced as it is used more efficiently and effectively. Discovery of new ideas and ways to solve problems, which is what innovation is all about, enhances the capability of a community or nation to grow economically and socially. This self-reinforcing capacity of a society to produce, reproduce, disseminate and share knowledge as a public good reaches a critical mass to propel it from a subsistence economy to a dynamic economy.

Innovation could be the creation of radically new or better products and processes. It could comprise the substitution of a cheaper material in an existing product, a cheaper way of producing it and better marketing of or supporting the product or service. It is both a value-adding and value-creating process. It consists of the adaptation of ideas or inventive activities into new and more efficient products or processes. Innovation is an interactive learning process that is rooted in a learning milieu (micro-foundation of innovation) formed by the co-operative efforts of various users, producers and other relevant social actors (Lundvall, 1992). Innovation is concerned with creating useful knowledge through the facilitation of adaptive social systems of innovation such that 'learning organisations' are 'able to adapt to knowledge' (de la Mothe, 2004). Innovation is about creating knowledge systems based on an algorithm of mixing local, acquired and learned knowledge. Although creativity plays an important role in promoting innovative activities, the ability to amass 'useful knowledge' (Mokyr, 2002) is critical. However, knowledge creation and transfer at present involve costs because they are proprietary (an imposed socially constructed concept and not a 'natural kind'), and almost all useful knowledge at the time of writing is being enclosed by large corporations and businesses through ratcheting up intellectual property rights claims.

It has been claimed that 'knowledge' has become the key to well-being and the alleviation of poverty. Rich nations create most of the technological knowledge and hence are the repository of almost all innovative activities and proprietorial technologies. They have the resources to invest in science and engineering education, cutting-edge R&D activities and the infrastructure to create knowledge out of knowledge, while poor nations are falling behind at an alarming pace as they lack such resources that are critical for innovation. As the rich nations create more income by riding the waves of innovation, they are able to create more jobs and economic opportunities and, consequently, accumulate further income and wealth. As innovations yield increased returns to scale, the gap between those nations and regions that are able to innovate and those that are unable to innovate widens.

Now, how do knowledge and innovation increase productivity, wealth and income? Analytically, the basic assumption is centred on the simple input–output relationship: that output is a function of independent input variables or factors of production (capital, labour, knowledge and so on). Productivity is determined by the efficiency with which the input factors are combined. The efficiency of such combination of the input factors – tangible (codified knowledge, machines, resources and people) as well as tacit (knowledge and skills) – is the real measure of technological progress. The key difference between traditional economy products and informational economy products is the difference in the cost function. Because of the increasing unit cost of production beyond certain economies of scale, traditional economy firms operate in the diminishing returns to scale mode. The reverse is the case for informational economy firms, which are operating mostly in the increasing returns mode because of the diminishing or almost negligible unit cost of (re)production. The increasing returns economy products have little congealed physical resources, but they have high knowledge intensity. Ironically, increasing returns to scale in industries or sectors was considered an anomaly, a crucial aspect of modern economic theory to which economists refused to pay attention because it did not conform to their inherited worldview (Arthur, 1994).

Increasing returns firms and businesses represent the integral part of informational capitalism – dominated by computers, information, communication, and bio-informatics technologies. In an increasing returns firm the marginal cost of production is initially relatively high compared to the marginal revenue (unit product price) because of the high production cost of the first unit such as software (an operating system such as Windows), hardware (a DRAM) or a bio-engineered (patented) drug. In general, the initial design, production, R&D and marketing costs of these products could be high, but this need not be the case in every instance. The higher initial production cost is not a norm because in many areas of the IT industry, such as open source software and Internet content development, product development cost is usually very low. However, the high upfront cost is involved only in the production of the first unit or first few units of a product or service. After that the replication or manufacturing cost (of a CD or software or drug) is rather negligible because of the steep decline in the marginal cost curve, which eventually merges with the supply curve. That is why new economy (increasing returns) high-tech products are heavy on know-how and light on resources, in contrast to traditional economy (diminishing returns) products and processes, which are high on congealed physical resources and low on know-how (Arthur, 1998).

The rate of profit for a knowledge economy firm is dependent upon the rate at which it can create new products based, again, on the rate at which new knowledge (ideas leading to inventions and innovations) is created and how well intellectual property rights claims can be established. Thus, in theory, all informational economy firms can reap high profits as long as they can create the demand (through timed obsolescence, shortening product cycles, rapid introduction of new versions and other marketing gimmicks) and stay in the market. The demand for informational economy products is further reinforced by lock-in and economy-wide network effects. That is, many high-tech products like PCs, operating systems, software, mobile phones and so on must be compatible with a network of users. The more users there are who use a product, the better and faster it captures the market share through the network effect. Market instability (markets tilt to favour a product that gets ahead), unpredictability (whether and for how long a product stays in the market) and multiple potential outcomes are the other hallmarks of increasing returns products (Arthur, 1998). This is the grim reality of digital capitalism that we need to grapple with to square grounded empirical reality in the global economy where one part (digital economy) appropriates more value than the other players. What this shows is that knowledge production is a self-reinforcing cycle that tends to disproportionately reward some and exclude others (Parayil, 2005).

The means by which knowledge economy (increasing returns) firms create value is through the intensification of intellectual property rights protection. Intellectual property protection does provide incentives for innovation and useful knowledge creation. However, the welfare gain to society due to intensifying IP protection yields diminishing returns on account of the monopolistic and rent-seeking behaviour of the firms. Like all good things, too much IP protection does not reward society (Drahos, 2003).[20] The intensification of IP benefits the owners of the innovations, while society, at large, suffers welfare loss due to the rent-seeking or monopolistic behaviour of knowledge economy firms that depend on patents, copyrights and other IPR regimes as their source of profit.

The critical economic development problem facing poor nations is the knowledge gap between them and rich nations. This knowledge deficit translates into a technology gap. The technology gap that was there historically facilitated the 'unequal exchange' between the advanced industrialised nations and the developing countries. According to Sachs (2001), the world is divided not by ideology, but by technology – a world of technology haves and have-nots. A small part of the globe made up of North America and parts of Europe and East Asia accounts for nearly all of the world's technological innovations and patent production. Much of the globe is technologically backward or excluded, being able neither to innovate nor to adopt and adapt new technologies (Sachs, 2001). This is the real face of digital divide (Parayil, 2005).

5. Conclusion and discussion

Recent trends in the global economy show a clear transition from an industrial era to an informational era. There has been a profound transformation of the economies of developed countries and parts of developing countries as a result of the rise of information capitalism. The increasing returns and competitiveness gains unleashed by the information revolution benefit only a few, while the majority of people who are left out of the IT revolution suffer a corresponding double loss due to the real digital divide and diminishing returns to their productive effort. While untold wealth was created for those connected to the winning camp of globalisation and the informational economy, it also spawned a skewed distribution of income and wealth in the world.

With the forced deregulation of the global economy, most states (especially the poor states) have been coerced into ceding their sovereignty to supra-national powers and unregulated finance capital. As a result, the capacity of the state to provide welfare measures to its citizens has been severely impaired, a situation similar to feudalism. The rise of the private sphere has severely eroded the public sphere in the information age, a situation detrimental to participative democracy, equality of opportunities and social welfare for the excluded.[21] These concepts had a powerful meaning during the long and drawn-out processes of democratisation and industrialisation. The reigning neo-liberal polity of the knowledge-based economy is poised to jettison these cherished social democratic reforms that were established through the struggles and tribulations of the previously excluded peoples of the industrial age. Despite the iniquities of the hard times faced by the working classes of industrial societies, the Industrial Revolution had a levelling effect on socio-economic and political differences (between classes) compared to its much more iniquitous agrarian–feudal predecessor as well its informational capitalist follower at the present time. Neo-liberal knowledge capitalism is evolving based on Margaret Thatcher's declaration that 'there is no such thing as society; there are only individual men and women and their families'.[22] Informational capitalism's primary objective, according to Dan Schiller (1999, p. 1), is 'paring unwanted state oversight and regulation of the economy to gain more unfettered freedom of action for private firms' so that markets could 'obey their presumed natural logic'.

Informational capitalism has spawned a new form of constant innovation, creative destruction, increasing returns and unimaginable wealth creation in the global economy. Information capitalism's expansion is more uneven and unpredictable than industrial capitalism. It has inaugurated a new international division of labour between countries and economic regions because of their differential competitiveness in an interdependent international market-place. The impact of globalisation of innovation and technological knowledge production on employment, work, leisure, income distribution and primary, secondary and tertiary production and so on is so unprecedented in human history that those who are still in the traditional (manufacture, extraction, agriculture, bulk-processing) modes of production and consumption can hardly cope with the changes brought about by the asymmetrical market relationships. Widening of the technological and wealth chasm between regions within countries and between countries is, ironically, occurring as the Information Revolution spreads across the globe.

While ICTs, like any other technology cluster, have the potential to benefit all, it is the unfair political economic context within which they are developed, deployed and diffused that needs to be reformed or better configured for equitable development. What is most urgent is to find ways to integrate informational economy with traditional economy in an equitable manner, such that the asymmetrical relationship between the two should be overcome by helping the latter to benefit from the advances in ICTs that played a key role in informational economy's shaping. It is meant, in this context, more than access to ICTs and bridging the so-called digital divide in an instrumental sense, the way the issue has been framed by many cyber-libertarian development experts. What developing countries like India should strive towards is to deploy and diffuse ICTs to enhance national competitiveness of all important sectors of the economy (such as agriculture, food-processing, manufacturing, transportation, banking, government services, education, retail services, tourism and healthcare) and bring about digital empowerment to improve the quality of life of all citizens, rather than concentrate merely on policies to reap short-term benefits from the foreign exchange windfall through the export of software products and services without any concern to its long-term developmental and distributional ramifications. Emphasis must be placed on the true opportunity cost of physical and human capital investment in the IT sector for boosting the export basket. The focus must shift from over-reliance on low value-added software and services export to the creation of new technologies suited to the large domestic market as well as high value-added branded products for overseas markets. ICTs should be diffused in all sectors of the economy, such that existing bottlenecks in the development and diffusion of useful knowledge, technology and information could be alleviated. By lowering communication, information-processing and search costs, SMEs could work in an environment with lower transaction costs and fewer entry barriers. It is not claimed with certainty, in a technological determinist sense, that ICT is a magic bullet that can solve all the entrenched social and economic problems. It is just that ICT can help if the right economic policies are in place, paying particular attention to its distributional and equity implications.

Notes

1. 'Knowledge' is an amorphous term, which is often tacit in nature and meant variously as idea, technological change, invention, innovation and so on.
2. The digital divide is one of the most misunderstood and misused concepts in the mushrooming discourse on information society. Since the literature on the digital divide is vast, no theoretical discussion of the concept will be carried out here. However, it may be pertinent to note that the very idea of the digital divide has been progressively constructed into a growth industry by both digital utopians and dystopians.
3. According to Brian Loader (1998), this is precisely due to many information society analysts' proclivity to view technological development as apolitical and outside the purview of the political economy. Mark Poster (1997) also questions the modernist attitude to view the Internet as resembling a 'hammer', a mere material configuration to perform some simple tasks. One must view it, he argues, as a socially and politically mediated object, rather than as a 'hammer', a thing without agency (although human agency is involved in constructing a hammer).
4. The neo-liberal position on globalisation professes the free movement of *all* factors of production, yet this is far from being the case for other factors, especially labour (except for some highly skilled IT and finance professionals).
5. This is despite Moore's Law that seems to dictate the inverse (supply-side) relationship between the price of IC chips and the speed of computing, which inevitably leads to price competition between producers of ICTs equipment. However, steep cuts in the prices of important input factors (computers and communication technologies) remain the key aspect of the information revolution.
6. See D'Costa (1999) for elaboration of this issue within the context of the rise of heavy industry such as steel in some developing countries.
7. Polemical works by Reich (1991), Rifkin (1995) and others on the impact of globalisation on labour must be taken in this context.
8. I am aware of the fact that capital is not cheap in the developing countries *vis-à-vis* developed countries due to higher rental cost consequent to higher real interest rate. The plant cost (part of fixed cost) is lower due to the availability of cheaper building materials, land, transportation and so on. Also, sophisticated plant machineries are still sourced and procured overseas for local manufacturing.
9. The rise of 'free trade' on a global scale with the rise of the WTO on the global stage in the 1990s is an important event here. Another important development was the new intellectual property rights (IPR) regime that evolved with Trade-Related Intellectual Property Rights (TRIPs) and General Agreement on Trade in Services (GATS). The nature of the manufacturing arrangement after the globalisation of production took many forms such as original equipment manufacture (OEM) contract whereby the developing country firm would assemble ICTs equipment for foreign MNCs and sell these under the MNC's original brand name.
10. The bursting of the dot-com bubble and the consequent decline in the for-

tunes of many silicone snake-oil peddlers (Stoll, 1995) should be seen as one of many other contradictory acts of information capitalism.

11. As mentioned earlier, Kuznets (1955) hypothesised an inverse relationship (U-shaped) between income inequality and GNP per capita.
12. It is important to point out that historically extreme inequality existed in countries such as Brazil, Mexico and India due to skewed distribution of land and other assets. However, industrialisation and modernisation had allowed some from the poor ranks to escape semi-feudal living conditions in the rural/agrarian sector.
13. We must note that most of the wealth and income accrued to the owners of the feudal estates as land rent. The number of people in the 'knowledge class' was very low compared to the serfs, who lived in abject poverty and deprivation. The 'knowledge class' was patronised by the landed aristocracy as the serfs were largely incapable or not allowed to seek the services of the knowledge class. I make no claim that this ratio was the same for all feudal societies. What is claimed is that on a global scale the average of this ratio would be about 1:100. It may have been lower in European feudal societies while higher in China. Although wage differential during feudal times may be difficult to compute, a proxy can be found in the distribution of wealth between the rich and the poor segments (for example the top and bottom deciles) in historical terms. For works on income and wealth distribution and studies on standards of living through the ages, see Burnette and Mokyr (1995), Lindert and Williamson (1995), Mokyr (1990), Maddison (1982, 2001 and 2003) and Brenner et al. (1991).
14. This is despite the caveat of the Kuznets curve in income distribution during the industrial era. What is important to note here is that the average income inequality across time between feudalism and industrialism would make the former more unequal than the latter.
15. This harrowing phase of British industrialism was eloquently portrayed by Dickens.
16. Again, this ratio is a global average. The US, since that has not a feudal past but only an agrarian past when industrialisation took off, was much more egalitarian than imperial Britain. However, with social democracy making headway in England and other European nations, the income inequality declined there in absolute and relative terms. The income differential may have been as low as 1:5 in some countries, while it could have been 1:20 in others. The claim of the global ratio of 1:10 should be taken as an average reference point of departure to compare with the past and the present. For excellent studies on income inequality and wage differentials since the Industrial Revolution in the US, England and other European countries, see Brenner et al. (1991) and Lindert and Williamson (1995). For other parts of the world, including the West, see Maddison (1982, 2001 and 2003) and Burnette and Mokyr (1995).
17. This is an improbable mathematical expression because it implies the absolute income of the knowledge class to be infinity ($1/0 = \infty$). The binary symbolism highlights the sharp contrast in income rather than the real income.
18. Arthur (1998) describes the unimaginable wealth created by high technology industries due to increasing returns as akin to casino gambling. Frank

and Cook (1995) vividly describe the growth of 'winner-take-all society' in the US since the onset of informational capitalism during the 1980s.

19. This is applicable to all the players who play the game fairly, with no coercion or imperialist tendencies present.
20. Plotting IP protection rate on the horizontal axis and welfare on the vertical axis will yield an inverted semi-circular curve. The ideal state of IP protection, yielding maximum welfare, will be in the top middle of the curve. Regions on the extreme left of the curve will lead to free-riding and low welfare; regions on the extreme right will lead to rent-seeking and low welfare (Drahos, 2003).
21. Not to be confused, I see civil society as part of the public sphere, and the public sphere in turn should not be confused with the state.
22. Quoted in Urry (2000, p. 5). One may agree with Urry that we do need to move beyond the concept of society in its current sociological sense. But Thatcher's project to demolish society was intended to destroy the welfare state and social democracy so as to let 'nature' take its course to usher in 'true' (*laissez faire*) capitalism, bearing its tooth and claw. This Thatcher–Reagan ideology is the corner-stone of neo-liberal globalisation.

Bibliography

Acharya, S. (2002) 'Macroeconomic management in the nineties', *Economic and Political Weekly*, 37(16): 1515–38.

Aghion, P., García-Penalosa, C. and Caroli, E. (1998) 'Inequality and Economic Growth', in Aghion, P. and Williamson, J. G. (eds) *Growth, Inequality, and Globalization: Theory, History, and Policy* (Cambridge: Cambridge University Press) pp. 5–102.

Ahluwalia, M. S. (1976) 'Inequality, Poverty and Development', *Journal of Development Economics*, 3: 307–42.

Alarcon, R. (1998) 'The Migrants of the Information Age: Foreign-born Engineers and Scientists and Regional Development in Silicon Valley', unpublished PhD dissertation, University of California, Berkeley.

Amin, A. and Wilkinson, F. (1999) 'Learning, Proximity and Industrial Performance: An Introduction', *Cambridge Journal of Economics*, 23: 121–5.

Amsden, A. H. (2001) *The Rise of 'the Rest': Challenges to the West from Late-Industrializing Economies* (New York: Oxford University Press).

——(1989) *Asia's Next Giant: South Korea and Late Industrialization* (New York: Oxford University Press).

Antonelli, C. (2003) 'The Digital Divide: Understanding the Economics of New Information and Communications Technology in the Global Economy', *Information Economics and Policy*, 15: 173–99.

——(1999) *The Microdynamics of Technological Change* (London: Routledge).

Arora, A. and Asundi, J. (1999) 'Quality Certification and the Economics of Contract Software Development: A Study of the Indian Software Industry', working paper 7260, National Bureau of Economic Research, Cambridge, MA.

——and Athreye, S. S. (2002) 'The Software Industry and India's Economic Development', *Information Economics and Policy*, 14: 253–73.

——, Arunachalam, V. S., Asundi, J. and Fernandes, R. (2001) 'The Indian Software Services Industry', *Research Policy*, 30: 1267–87.

Arrow, K. J. (1962) 'Economic Welfare and the Allocation of Resources for Invention', in *The Rate and Direction of Inventive Activity* (Princeton: NBER, Princeton University Press) pp. 609–26.

Arthur, W. B. (1998) 'Increasing Returns and the New World of Business', in Neef, D. (ed.) *The Knowledge Economy* (Boston: Butterworth-Heinemann) pp. 75–85.

——(1994) *Increasing Returns and Path Dependence in the Economy* (Ann Arbor, MI: University of Michigan Press).

Arun, S. and Arun, T. (2002) 'ICTs, Gender and Development: Women in Software Production in Kerala', *Journal of International Development*, 14: 39–50.

Autor, D. H., Katz, L. F. and Krueger, A. B. (1998) 'Computing Inequality: Have Computers Changed the Labor Market?', *Quarterly Journal of Economics*, 113: 1169–1213.

Balaji, V., Rajamohan, K. G., Rajashekhara P. R. and Kumaran, S. S. (2001) 'Towards Knowledge Systems for Sustainable Food Security: The Information Village Experiment in Pondicherry', in Das, M. R. (ed) *Proceedings of the Theme*

Session: Information Technology for Development – a Millennium Perspective, XIII Kerala Science Congress, Thrissur, India, 29–31 January.
Banerjee, A. V. and Duflo, E. (2000) 'Reputation Effects and the Limits of Contracting: A Study of the Indian Software Industry', *Quarterly Journal of Economics*, 115: 989–1017.
Barboza, D. (2004) 'An Unknown Giant Flexes its Muscles: Lenovo's Bid for IBM's PC Unit is Another Step onto Global Stage', *New York Times*, 4 December, B1, B3.
Beije, P. (1998) *Technological Change in the Modern Economy: Basic Topics and New Developments* (Cheltenham: Edward Elgar).
Bell, D. (1973) *The Coming of Post-Industrial Society: A Venture in Social Forecasting* (New York: Basic Books).
Berman, E., Bound, J. and Griliches, Z. (1994) 'Changes in the Demand for Skilled Labor within U.S. Manufacturing Industries: Evidence from the Annual Survey of Manufacturers', *Quarterly Journal of Economics*, 109: 367–97.
——, Bound, J. and Machin, S. (1998) 'Implications of Skill-Biased Technological Change: International Evidence', *Quarterly Journal of Economics*, 113: 1245–79.
Bhagwati, J. N. (1984) 'Why are Services Cheaper in Poor Countries?' Economic Journal, 94: 279–85.
Bhatnagar, S. and Schware R. (eds) (2000) *Information and Communication Technology in Development: Cases from India* (New Delhi: Sage Publications).
Bhuyan, R. (2002) 'Chipping in', *Data Quest*, at http://www.dqindia.com/content/search/showarticle.asp?arid=39564&way=search, accessed 26 March 2004.
Borger, J. (2000) 'Workers' Rights Abused in US', *The Guardian*, 30 August.
Braithwaite, J. and Drahos, P. (2000) *Global Business Regulation* (Cambridge: Cambridge University Press).
Brenner, Y. S., Kaelble, H. and Thomas, M. (1991) *Income Distribution in Historical Perspective* (Cambridge: Cambridge University Press).
Bresnahan, T. F. (1997) 'Computerization and Wage Dispersion: An Analytical Reinterpretation' mimeo, Stanford University.
Brooks, F. P. (1995) *The Mythical Man-Month: Essays on Software Engineering* (New York: Addison Wesley Longman).
Burnette, J. and Mokyr, J. (1995) 'The Standard of Living Through the Ages', in Simon, J. L. (ed.), *The State of Humanity* (Oxford: Blackwell) pp. 135–48.
Burton-Jones, J. (1999) *Knowledge Capitalism: Business, Work, and Learning in the New Economy* (New York: Oxford University Press).
Business Standard (2004) 'Rising Re May Hit IT Firms' Q3 Earnings', 4 December, www.business-standard.com, accessed 7 December 2004.
Cantwell, J. and Santangelo, G. D. (1999) 'The Frontier of International Technology Networks: Sourcing Abroad the Most Highly Tacit Capabilities', *Information Economics and Policy*, 11: 101–203.
Caselli, F. (1999) 'Technological Revolutions', *American Economic Review*, 89: 78–102.
Castells, M. (1996) *The Rise of the Network Society, Vol. I The Information Age: Economy, Society and Culture* (Oxford: Blackwell).
——(1993) 'The Informational Economy and the New International Division of Labor', in Carnoy, M., Castells, M., Cohen, S. S. and Cardoso, F. H. (eds), *The*

New Global Economy in the Information Age: Reflections on our Changing World (University Park, PA: Pennsylvania State University Press) pp. 16–43.

——and Hall, P. (1994) *Technopoles of the World: the Making of Twenty-First-Century Industrial Complexes* (London: Routledge).

Central Statistical Organisation (1999) *Statistical Abstract of India 1998*, Department of Statistics and Programme Implementation, Government of India, New Delhi.

Chandrasekhar, C. P. (2003) 'The Diffusion of Information Technology: The Indian Experience', *Social Scientist*, 31(7–8): 42–85.

——and Ghosh, J. (2001) 'Information and Communication Technologies and Health in Low Income Countries: The Potential and the Constraints', *Bulletin of the World Health Organisation*, 79(9): 850–9.

Chibber, V. (2003) *Locked in Place: State-Building and Late Industrialization in India* (Princeton: Princeton University Press).

Cohen, D. (1998) *The Wealth of the World and the Poverty of Nations* (Cambridge, MA: MIT Press).

Cohen, Y. and Haberfeld, Y. (1993) 'Temporary Help Service Workers: Employment Characteristics and Wage Determination', *Industrial Relations*, 32: 272–87.

Computer Today (2001) 'Study Reinforces Recession Fears in US', *Computer Today*, 16–23 July: 25.

Corden, W. M. (1984) 'Booming Sector and Dutch Disease Economics: Survey and Consolidation', *Oxford Economic Papers*, 35: 359–80.

——and Neary, J. P. (1982) 'Booming Sector and De-industrialisation in a Small Open Economy', *Economic Journal*, 92: 825–48.

Correa, C. M. (1996) 'Strategies for Software Exports from Developing Countries', *World Development*, 24: 171–82.

Cresswell, E. (2000) 'Contracting Seen as Way of Entering IT Industry', *Australian IT*, 30 June.

Davis, A. (1993) 'Software Lemmingineering', *IEEE Software*, 10(5): 79–81, 84.

Davis, S. J., Haltiwanger, J., Katz, L. E. and Topel, R. (1991) 'Wage Dispersion between and within U.S. Manufacturing Plants, 1963–86', *Brookings Papers on Economic Activity: Microeconomics*, 1991: 115–80.

D'Costa, A. P. (2005) *The Long March to Capitalism: Embourgeoisment, Internationalization, and Industrial Transformation in India* (Basingstoke: Palgrave Macmillan).

——(2004a) 'Flexible Institutions for Mass Production Goals: Economic Governance in the Indian Automotive Industry', *Industrial and Corporate Change* 13(2): 335–67.

——(2004b) 'Globalization, Development, and Mobility of Technical Talent: India and Japan in Comparative Perspectives', research paper no. 2004/62, United Nations University, World Institute of Development Economics Research.

——(2003a) 'Catching Up and Falling Behind: Inequality, IT, and the Asian Diaspora', in Ho, K. C. et al. (eds), *Asia Encounters the Internet* (London: Routledge), pp. 44–66.

——(2003b) 'The Indian Software Industry in the Global Division of Labor', in D'Costa, A. P. and Sridharan, E. (eds), *India in the Global Software Industry: Innovation, Firm Strategies and Development* (Basingstoke: Palgrave Macmillan), pp. 1–26.

——(2003c) 'Uneven and Combined Development: Understanding India's Software Exports,' *World Development*, 31(1): 211–26.
——(2002a) 'Export Growth and Path-Dependence: The Locking-In of Innovations in the Software Industry', *Science, Technology and Society*, 7: 51–89.
——(2002b) 'Software Outsourcing and Policy Implications: An Indian Perspective', *International Journal of Technology Management*, 24(7/8): 705–23.
——(2002c) 'Technological Leapfrogging: The Software Challenge in India', in Conceição, P. et al., *Knowledge for Inclusive Development* (New York: Quorum Books), pp. 183–99.
——(2001) 'Creeping Innovative Capability in the Indian Software Industry: Learning Strategies of Firms', presented at the Workshop on Indian Software Industry in a Global Context, Indian Institute of Management, Ahmedabad, India, 14–25 August.
——(1999) 'The Restructuring of the Indian Automobile Industry: Indian State and Japanese Capital', in Buckley, P. and Ghauri, P. N. (eds), *The Global Challenges for Multinational Enterprises: Managing Increasing Interdependence* (Amsterdam & New York: Pergamon), pp. 189–224.
Deccan Herald (2002) 'IT Sector May Achieve $50 bn Exports by 2008, says Study', *Deccan Herald*, 11 June.
Dedrik, J. and Kraemer, K. L. (1998) *Asia's Computer Challenge: Threat or Opportunity for the United States & the World?* (New York: Oxford University Press).
de la Mothe, J. (2004) 'The Institutional Governance of Technology, Society and Innovation', *Technology in Society*, 26: 523–36.
Department of Agriculture (2002) 'Uzhavar Uthaviyakam: Oru Arizhakam', Tamil booklet, Pondicherry.
Desai, A. V. (2001) 'The Peril and the Promise: Broader Implications of the Indian Presence in Information Technologies', Workshop on the Indian Software Industry in a Global Context, Indian Institute of Management, Ahmedabad, India, 14–25 August.
Dicken, P. (2003) *Global Shift: Reshaping the Global Economic Map in the 21st Century*, 4th edition (New York: Guilford Press).
Dirlik, A. (1999) 'Place-based Imagination: Globalism and the Politics of Place', *Review, A Journal of the Fernand Braudel Center*, 22(2): 151–87.
Dornbusch, R. (1992) 'The Case for Trade Liberalization in Developing Countries', *Journal of Economic Perspectives*, 6: 69–85.
Drahos, P. (2003) 'The Global Intellectual Property Ratchet in the Information Age: Consequences and Costs', paper presented at the Information & Communications Management Programme, National University of Singapore, 3 December.
——(1995) 'Information Feudalism in the Information Society', *The Information Society*, 11: 209–22.
——and Braithwaite, J. (2002) *Information Feudalism: Who Owns the Knowledge Economy?* (New York: The New Press).
Drèze, J. and Sen, A. (2002) *India: Development and Participation* (New Delhi: Oxford University Press).
Drucker, P. (1993) *Post-Capitalist Society* (New York: HarperBusiness).
——(1969) *The Age of Discontinuity* (New York: Harper & Row).
Dugger, C. (2000) 'Connecting Rural India to the World', *New York Times*,

28 May at http://partners.nytimes.com/library/tech/00/05/biztech/articles/28india.html, accessed 15 November 2004.

Dunn, D. and Yamashita, K. (2003) 'Microcapitalism and the Megacorporation', *Harvard Business Review*, August: 46–54.

Dunn, L. (2001) 'Landscape shifts for laid-off foreigners', *CBS Marketwatch*, at http://www.marketwatch.com/, accessed 31 August 2001.

Dyer-Witheford, N. (1999) *Cyber-Marx: Cycles and Circuits of Struggle in High-Technology Capitalism* (Urbana and Chicago: University of Illinois Press).

The Economist, (2001) 'Mobile Phones in India: Another Kind of Network', *The Economist*, 3 March: 63.

Egan, E. A. (1997) *The Spatial Dynamics of the U.S. Computer Software Industry*, unpublished PhD dissertation, University of California, Berkeley.

Eggleston, K., Jensen, R. and Zeckhauser, R. (2002) 'Information and Communication Technologies, Markets, and Economic Development' in Kirkman, G. S., Cornelius, P. K., Sachs, J. D. and Schwab, K. (eds), *The Global Information Technology Report 2001–2002: Readiness for the Networked World* (New York: Oxford University Press), pp. 62–75.

Eischen, K. (2000) 'National Legacies, Software Technology Clusters and Institutional Innovation: The Dichotomies of Regional Development in Andhra Pradesh, India', working paper 2000–2002, Center for Global, International and Regional Studies, University of California, Santa Cruz.

Evans, P. B. (1995) *Embedded Autonomy: States and Industrial Transformation* (Princeton: Princeton University Press).

——(1992) 'Indian Informatics in the 1980's: Changing Character of State Involvement', *World Development*, 21: 1–18.

Feenstra, R. C. (1998) 'Integration of Trade and Disintegration of Production in the Global Economy', *Journal of Economic Perspectives*, 12(4): 31–50.

——and Hanson, G. H. (1996) 'Globalisation, Outsourcing and Wage Inequality', *American Economic Review*, 86: 240–5.

Financial Express, (2004) 'Guard Against Complacency in IT Sector: Study', *Financial Express*, 3 December, at www.financialexpress.com, accessed 7 December 2004.

Financial Times, (1995) 'Review of Information Technology: India's software industry', *Financial Times* I–VI, 6 December.

Frank, R. H. and Cook, P. J. (1995) *The Winner-Take-All-Society* (New York: The Free Press).

Friedman, T. (2000) *The Lexus and the Olive Tree* (New York: HarperCollins).

Fröbel, F., Heinrichs, J. and Kreye, O. (1981) *The New International Division of Labour* (Cambridge: Cambridge University Press).

Gibbs, W. W. (1994) 'Software's chronic crisis', *Scientific American*, 271(3): 86–95.

Gordon, D. M. (1996) *Fat and Mean: The Corporate Squeeze of Working Americans and the Myth of Managerial 'Downsizing'* (New York: Martin Kessler Books).

Gottfried, H. (1991) 'Mechanisms of control in the temporary help service industry', *Sociological Forum*, 6(4): 699–713.

Government of Andhra Pradesh (2000) *AP First: Information Technology Policy* (Hyderabad).

Government of India (2004) *Economic Survey* (New Delhi: Ministry of Finance), at http://indiabudget.nic.in/, accessed 20 November 2004.

Green, F., Dickerson, A. and Arbache, J. S. (2001) 'A Picture of Wage Inequality and the Allocation of Labor Through a Period of Trade Liberalization: The Case of Brazil', *World Development*, 29: 1923–39.

Guha, B. (2003) 'IT: Deconstructing the Bust that Followed the Boom', *Economic and Political Weekly*, 14 June, at http://epw.org.in/showArticles.php?root=2003&leaf=06&filename=5924filetyp . . . , accessed 20 June 2003.

Guo, J. and Planting, M. A. (2000) 'Using Input–Output Analysis to measure YC Economic Structural Change over a 24-year period', paper presented at the International Conference on Input–Output Techniques, Macerata, Italy, 21–25 August; Washington DC, Bureau of Economic Analysis, US Department of Commerce, at http://www.bea.doc.gov/bea/papers/strucV7all.pdf, accessed June 2004.

Hamel, G. and Prahalad, C. K. (1994) *Competing for the Future* (Boston: Harvard Business School Press).

Harding, E. U. (1989) 'India: After IBM's Exit, an Industry Arose', *Software Magazine*, 9(14): 48–54.

Hari, P. (2003) 'Embedded technology: the crossover', *Bussiness World*, 14 April: 28–32.

——and Anand, M. (2002) 'Chip's of the Block,' *Business World*, 18 March: 34–41.

Hart, J. A. and Kim, S. (2002) 'Explaining the Resurgence of U.S. Competitiveness: The Rise of Wintelism', *The Information Society*, 18: 1–12.

Harvey, D. (1989) *The Condition of Post Modernity: An Enquiry into the Origins of Cultural Change* (Oxford: Blackwell).

Heeks, R. (2002) 'i-Development not e-Development: Special Issue on ICTs and Development', *Journal of International Development*, 14: 1–11.

——(1996) *India's Software Industry: State Policy, Liberalisation and Industrial Development* (New Delhi: Sage Publications).

Held, D., McGrew, A., Goldblatt, D. and Perraton, J. (1999) *The Global Transformations: Politics, Economics and Culture* (Stanford: Stanford University Press).

Henderson, J. (1989) *The Globalisation of High Technology Production: Society, Space and Semiconductors in the Restructuring of the Modern World* (London: Routledge).

——, Dicken, P., Hess, M., Coe, N. and Yeung, H. W. C. (2002) 'Global Production Networks and the Analysis of Economic Development', *Review of International Political Economy*, 9(3): 436–64.

The Hindu, (2004) 'Pro-rich Image, Pessimism about Reforms did Tamil Naidu in', *The Hindu*, 12 May at http://www.hinduonnet.com/2004/05/12/stories/2004051207311100.htm.

Hirschman, A. (1970) *Exit, Voice, and Loyalty* (Cambridge, MA: Harvard University Press).

Holmstrom, B. and Roberts, J. (1998) 'The Boundaries of the Firm Revisited', *Journal of Economic Perspectives*, 12: 73–94.

IDRC [International Development Research Center], (1999) 'Success Stories of Rural ICTs in a Developing Country: Report of the PANAsia Telecenter Learning & Evaluation Group's mission to India, involving visits to the Foundation of Occupational Development and the M. S. Swaminathan research foundation', at http://www.msssrf.org, retrieved on 16 November 2002.

IMF, (2001) *World Economic Outlook: The Information Technology Revolution* (Washington, DC: IMF).

InfoTech Consulting, (1992) India's Software and Services: Export Potential and Strategies, report submitted in fulfilment of contract with the Government of India, Department of Electronics, New Delhi (under World Bank Japan grant #2612).

INS [U.S. Immigration and Naturalization Service], (2000) 'Characteristics of Specialty Occupation Workers (H-1B): May 1998 to July 1999', February.

Jain, S. (2002) 'Not all IT's cracked up to be', *Indian Express*, 26 January.

James, J. (2001) 'Information Technology, Cumulative Causation and Patterns of Globalization in the Third World', *Review of International Political Economy*, 8: 147–62.

Jhunjhunwala, A. (2001) 'Looking Beyond NTP 99', *3iNetwork*.

Jishnu, L. M., Anand, A. and Hari, P. (2001) 'India's Great Digital Hope', *Business World*, 21 May: 28–41.

Jones, C. (1998) 'Bad Days for Software', *IEEE Spectrum*, 35(9): 47–52.

Joseph, K. J. (2002) 'Growth of ICT and ICT for Development: Realities of the Myths of the Indian Experience', discussion paper no. 2002/78 (Helsinki: UNU/WIDER)

——(1997) *Industry under Economic Liberalization: The Case of Indian Electronics* (New Delhi: Sage Publications).

——and Harilal, K. N. (2001) 'India's IT Export Boom: Challenges Ahead', working paper no. 317, Centre for Development Studies, Thiruvananthapuram, India, at http://www.cds.edu/download_files/wp317.pdf, accessed 20 February 2005.

——and Parayil, G. (2005) 'ASEAN India Cooperation in Information Communication Technology and Enabled Services: Issues and Prospects', in Sen, R., Asher, M. and Kumar, N. (eds), *ASEAN-India Partnership: Meeting the Challenges of Integrating World* (Singapore: ISEAS, in press).

Kalleberg, A. L. (2000) 'Nonstandard Employment Relations: Part-time, Temporary and Contract Work', *Annual Review of Sociology*, 26: 341–65.

Kamas, L. (1986) 'Dutch Disease Economics and the Colombian Export Boom', *World Development*, 14: 1177–98.

Kaplan, L. (2000) 'A New Frontier for the MNC', at http://www.deloitte.com, accessed 1 September 2001.

Kapoor, S. (2001) 'M. S Swaminathan: Brain Food for the Masses', *Asiaweek*, 29 June: 35–6.

Kattuman, P. and Iyer, K. (2001), 'Human Capital in the Move Up the Value Chain: The Case of the Indian Software and Services Industry', in Kagami, M. and Tsuji, M. (eds), *The 'IT' Revolution and Developing Countries: Latecomer Advantage?* (Tokyo: Institute of Developing Economies, Japan External Trade Organization) pp. 208–27.

Kaushik, P. D. and Singh, N. (2004) 'Information Technology and Broad-Based Development: Preliminary Lessons from North India', *World Development*, 32: 591–607.

Keeble, L. and Loader, B. D. (eds), (2001) *Community Informatics: Shaping Computer Mediated Social Relations* (London and New York: Routledge).

Keniston, K. (2002) 'Grass Roots ICT Projects in India: Some Preliminary Hypothesis', *ASCI Journal of Management*, 31(1&2), at http://web.mit.edu/~kken/ Public/PAPERS/ASCI_Journal_Intro__ASCI_ version_.html, accessed 15 November 2004.

Keuffel, W. (1991) 'House of Structure', *UNIX Review*, 9(2): 28–36.
Khadria, B. (2001) 'Shifting Paradigms of Globalization: The Twenty-First Century Transition towards Generics in Skilled Migration from India', *International Migration*, 39(5): 45–71.
Khanna, R. (2001) 'TARAhaat at a Glance', *Development Alternatives Newsletter*, 11(7): 7–8.
Kirby, C. (2001) 'Documenting an American Dream', *San Francisco Chronicle*, 1 July.
Kitchenham, B. and Pfleeger, S. W. (1996) 'Software Quality: The Elusive Target', *IEEE Software*, 13(1): 361–84.
Koch, K. (1998) High-tech labor shortage', *CQ Researcher*, 8(16): 361–84.
Kraemer, K. L. and Derick, J. (2001) 'Information Technology and Economic Development: Results and Policy Implications of Cross-Country Studies', in Pohjola, M. (ed.), *Information Technology, Productivity and Economic Growth* (Oxford: Oxford University Press) pp. 257–80.
Kraft, P. (1977) *Programmers and Managers: The Routinization of Computer Programming in the United States* (New York: Springer-Verlag).
Krishnadas, K. C. (2003a) 'India's Design Centers Buck Economy's Trend', at http://www.eetimes.com/article/showArticle.jhtml?articleId=18308505, accessed 20 February 2005.
——(2003b) 'Indian EEs show you can go home again', at http://www.eetimes.com/article/showArticle.jhtml?articleId=18309505, accessed 20 February 2005.
——(2000) 'India Readies Laws to Protect IC Designs', at http://www.eedesign.com/article/showArticle.jhtml?articleId=17405970, accessed 20 February 2005.
Krishnakumar, A. (2001) 'Changing Rural Lives', *Frontline*, 18(24), at http://www.flonnet.com/fl1824/18240990.htm, accessed 18 November 2001.
Krueger, A. (1993) 'How Computers Have Changed the Wage Structure: Evidence from Microdata, 1984–89', *Quarterly Journal of Economics*, 108: 33–59.
Krugman, P. (1995) 'Technological Change in International Trade,' in Stoneman, P. (ed.) *Handbook of the Economics of Innovation and Technological Change* (Oxford: Blackwell), pp. 342–65.
Kumar, A. (2000) 'Computerisation of *Mandal* Revenue Offices in Andhra Pradesh: Integrated Certificate Application', in Bhatnagar and Schware, pp. 94–104.
Kumar, N. (2001) 'Indian Software Industry Development: International and National Perspective', *Economic and Political Weekly*, 36(10): 4278–90.
——(2000) 'Small Information Technology Services, Employment and Entrepreneurship Development: Some Explorations into Indian Experience', *Indian Journal of Labour Economics*, 43(4): 935–48.
——and Joseph, K. J. (2004) 'National Innovation Systems and India's IT Capability: Are there any Lessons for ASEAN New Comers?', RIS discussion paper no. 72/2004 (New Delhi: Research and Information System for Non-Aligned and other Developing Countries).
Kumar, V. R. (2003) 'Touching base', at http://www.blonnet.com/ew/2003/07/02/stories/2003070200070100.htm, accessed 20 February 2005.
Kuznets, S. (1955) 'Economic Growth and Income Inequality', *American Economic Review*, 45: 1–28.

Lakha, S. (1999) 'The New International Division of Labour and the Indian Computer Software Industry', in Bryson, J. et al. (eds), *The Economic Geography Reader: Producing and Consuming Global Capitalism* (Chichester: John Wiley) pp. 148–55.

——(1990) 'Growth of computer software industry in India', *Economic and Political Weekly*, 25(1): 49–56.

Layton, E. T. (1974) 'Technology as Knowledge', *Technology and Culture*, 15: 31–41.

Lee, E. A. (2000) 'What's Ahead for Embedded Software?', *IEEE Computer*, 33(9): 18–26.

Lema, R. and Hesbjerg, B. (2003) *The Virtual Extension: A Search for Collective Efficiency in the Software Cluster in Bangalore* (Roskilde: Public Administration and Public Economics & International Development Studies, Roskilde University).

Lindert, P. H. and Williamson, J. G. (1995) 'The Long-Term Course of American Inequality: 1647–1969', in Simon, J. L. (ed.), *The State of Humanity* (Oxford: Blackwell) pp. 188–95.

Lindsey, B. (2004) 'Job Losses and Trade: A Reality Check' (Washington: Centre for Trade Policy Studies, Cato Institute, 17 March), at http://www.freetrade.org/pubs/briefs/tbp-019.pdf, accessed December 2004.

Loader, B. (1998) 'Cyberspace Divide', in Loader, B. (ed.), *Cyberspace Divide: Equality, Agency and Policy in the Information Society* (London: Routledge), pp. 3–16.

Lubeck, P. and Eischen, K. (1998) 'Silicon Islands and Silicon "Valleys": Rethinking Mexican Regional Development Strategies in an Era of Globalization', in Alvarez, A., Castillo, P., Klahn, N. and Manchon, F. (eds), *Las Nuevas Fronteras del Siglo XXI: Dimensiones Culturales, Politicas y Socioeconomicas de las Relaciones Mexico-Estados Unidos* (Mexico City: UNAM/UAM), pp. 659–82.

Lubman, S. (2000) 'Labor Contractor of All Stripes Abound in Bay Area and the Nation', *San Jose Mercury News*, 19 November.

Lucas, H. C. and Sylla, R. (2003) 'The Global Impact of the Internet: Widening the Economic Gap Between Wealthy and Poor Nations?', *Prometheus*, 21: 3–22.

Lum, S. K. S., Moyer, B. C. and Yuskagave, R. E. (2000) 'Improved Estimates of Gross Product by Industry for 1947–98', in *Survey of Current Business*, June, pp. 214–54, at http://www.bea.gov/bea/articles/National/NIPAREL/2000/0600gpi.pdf, accessed June 2004.

Lundvall, B.-Å. (ed.) (1992) *National Systems of Innovation: Towards a Theory of Innovation and Interactive Learning* (London: Pinter).

Maddison, A. (2003) *The World Economy: Historical Statistics* (Paris: OECD).

——(2001) *The World Economy: A Millennial Perspective* (Paris: OECD).

——(1982) *Phases of Capitalist Development* (Oxford: Oxford University Press).

Malecki, E. and Oinas, P. (eds) (1999) *Making Connections: Technological Learning and Regional Economic Change* (Brookfield: Ashgate).

Mandel, E. (1978) *The Second Slump: A Marxist Analysis of Recession in the Seventies* (translated from German by J. Rothschild), (London: NLB).

——(1975) *Late Capitalism* (translated from German by J. De Bres), (London: NLB).

Mansell, R. (2002) 'From Digital Divides to Digital Entitlements in Knowledge Society', *Current Sociology*, 50: 407–26.

——and Wehn, U. (eds) (1998) *Knowledge Societies: Information Technology for Sustainable Development* (New York: United Nations Commission on Science and Technology for Development and Oxford University Press).
Mansfield, E. (1985) *Micro-Economics: Theory and Applications* (New York: W. W. Norton).
Mariani, M. (1999) 'Next to Production or to Technological Clusters? The Economics and Management of R&D Location', unpublished paper, University of Maastricht.
Martin, G. and Schirrmeister, F. (2002) 'A design chain for embedded systems', *IEEE Computer*, 35(3): 100–3.
Maskell, P. and Malmberg, A. (1999) 'Localised Learning and Industrial Competitiveness', *Cambridge Journal of Economics*, 23: 167–85.
Mathews, J. A. and Cho, D. S. (2000) *Tiger Technology: The Creation of a Semiconductor Industry in East Asia* (Cambridge: Cambridge University Press).
Mayall, A. (1991) 'Spaghetti Systems', *Which Computer?*, 14(1): 40.
McKinsey Global Institute (2003) *Offshoring: Is it a win-win game?* (San Francisco: McKinsey Global Institute), at http://www.mckinsey.com/mgi/reports/pdfs/offshore/Offshoring_MGI_Perspective.pdf, accessed July 2004.
Menon, R. K. (2003) 'IC Design Houses Swing into High Gear', at http://economictimes.indiatimes.com/articleshow/331891.cms, accessed 20 February 2005.
Mohan, R. (1998) 'Products vs. Services: The Argument Continues', *Business Line*, 17 December: 7.
Mokyr, J. (2002) *The Gifts of Athena: Historical Origins of the Knowledge Economy* (Princeton: Princeton University Press).
——(1990) *The Lever of Riches: Technological Creativity and Economic Progress* (New York: Oxford University Press).
Moreira, M. M. (1995) *Industrialization, Trade and Market Failures: The Role of Government Intervention in Brazil and South Korea* (Basingstoke, Macmillan Press).
Morris-Suzuki, T. (1997) 'Capitalism in the Computer Age and Afterword', in Davis, J., Hirschl, T. A. and Stack, M. (eds), *Cutting edge: technology, information and social revolution* (London: Verso), pp. 57–72.
Mowery, D. C. and Oxley, J. E. (1995) 'Inward Technology Transfer and Competitiveness: The Role of National Innovation System', *Cambridge Journal of Economics*, 19: 67–93.
MSSRF (2002a) *Annual Report 2000–2001* (Chennai: M. S. Swaminathan Research Foundation), at http://www.mssrf.org, retrieved 17 November 2002.
——(2002b) *Annual Report 2001–2002* (Chennai: M. S. Swaminathan Research Foundation), at http://www.mssrf.org, retrieved 17 November 2002.
Mulcaster, G. (2001) 'Headhunters Join Forces', *IT Jobs* (Australia), 6 November.
Nag, K. (2001) 'IT jobs are here again!', *Times of India*, Hyderabad, 9 July.
Nair, C. (1999) 'Cadence inks Multi-Year Pacts with Indian Design Houses', at http://www.eetimes.com/article/shiwArticle.jhtml?articleId=18303272, accessed 20 February 2005.
Nair, J. (2000) 'Singapore is not Bangalore's Destiny', *Economic and Political Weekly*, 35(18): 1512–14.
Narula, R. (1999) 'In-House R&D, Outsourcing or Alliances? Some Strategic and Economic Considerations', unpublished paper, University of Oslo.

NASDAQ (2001) 'About NASDAQ', at http://www.nasdaq.com/about/about_nasdaq.stm, accessed December 2001.

NASSCOM (2004) *The Software Industry in India: A Strategic Review* (New Delhi: National Association of Software and Service Companies).

——(2003) *The Impact of Global Sourcing on the US Economy 2003–2010* (New Delhi: National Association of Software and Service Companies).

——(2002) *The IT Industry in India: Strategic Review 2002* (New Delhi: National Association of Software and Service Companies).

——(2001) *Indian IT and Software Services Directory, 2001* (New Delhi: National Association of Software and Service Companies).

——(1999) *The Software Industry in India: A Strategic Review* (New Delhi: National Association of Software and Service Companies).

——(1993) *Indian Software Directory, 1993–94* (New Delhi: National Association of Software and Service Companies).

——(1992) *Indian Software Directory, 1992* (New Delhi: National Association of Software and Service Companies).

Nayyar, D. (1988) 'The Political Economy of International Trade in Services', *Cambridge Journal of Economics*, 12: 279–98.

Neary, J. P. (1978) 'Short Run Capital Specificity in the Pure Theory of International Trade', *Economic Journal*, 88: 488–510.

Nolan, P. (2001) 'The changing nature of employment', working paper, Australian Centre for Industrial Relations Research and Training, Sydney).

NTIT&SD (1998) *IT Action Plan* (New Delhi: National Taskforce on Information Technology and Software Development), at http://it-taskforce.nic.in, accessed 20 February 2001.

OECD (2002) *Information Technology Outlook: ICTs and the Information Economy* (Paris: Organisation for Economic Cooperation and Development).

——(2000a) *Information Technology Outlook: ICTs, E-Commerce and the Information Economy* (Paris: Organisation for Economic Cooperation and Development).

——(2000b) *A New Economy? The Changing Role of Innovation and Information Technology in Growth* (Paris: Organisation for Economic Cooperation and Development).

——(2000c) *The Service Economy* (Paris: Organisation for Economic Cooperation and Development).

Okhi, H. (2001) 'International Division of Labor in East Asia's IT Industry', in Kagami, M. and Tsuji, M. (eds), *The 'IT' Revolution and Developing Countries: Latecomer Advantage?* (Tokyo: Institute of Developing Economies, Japan External Trade Organization), pp. 63–91.

Oman, C. (ed.) (1996) *Policy Reform in India* (Paris: Organisation for Economic Cooperation and Development).

Pais, J. (2003) Production Units and the Workforce in the Urban Informal Sector: A Case Study from Mumbai, unpublished PhD thesis, Indira Gandhi Institute of Development Research, Mumbai.

Parayil, G. (2005) 'The Digital Divide and Increasing Returns: Contradictions of Informational Capitalism', *The Information Society*, 21: 41–51.

——(ed.) (2000) *Kerala – The Development Experience: Reflections on Sustainability and Replicability* (London: Zed Books).

——(1999) *Conceptualizing Technological Change* (Lanham, MD: Rowman & Littlefields).

Parfitt, T. (2002) *The End of Development: Modernity, Post Modernity and Development* (London: Pluto Press).

Parrinello, S. (2004) 'The Service Economy Revisited', *Structural Change and Economic Dynamics*, 15: 381–400.

Parthasarathi, A. and Joseph, K. J. (2002) 'Limits to Innovation with Strong Export Orientation: The Experience of India's Information Communication Technology Sector', *Science, Technology and Society*, 7: 13–49.

Parthasarathy, B. (2004) 'India's Silicon Valley or Silicon Valley's India?: Socially Embedding the Computer Software Industry in Bangalore', *International Journal of Urban and Regional Research*, 28: 664–85.

Patel, I. (2002) 'Information and Communication Technology and Distance Adult Literacy Education in India', working paper no. 166, Institute of Rural Management, Anand, Gujarat.

Paulk, M. C. (1995) 'How ISO 9001 compares with CMM. IEEE Software', 12: 74–83.

Peterson, C., Sandell, V. and Lawlor, A. (2001) *What Works: TARAhaat's Portal for Rural India* (Washington, DC: The World Resources Institute), at http://www.digitaldividend.org/pdf/tarahaat.pdf, retrieved 18 November 2002.

Petras, J. (1999a) 'Globalization: A Critical Analysis', *Journal of Contemporary Asia*, 29(1): 3–37.

——(1999b) 'NGOs in the Service of Imperialism', *Journal of Contemporary Asia*, 29(4): 429–40.

Picketty, T. and Saez, E. (2003) 'Income Inequality in the United States, 1913–1998', *Quarterly Journal of Economics*, 118: 1–39.

Pieterse, N. (2000) 'Trends in Development Theory', in Palan, R. (ed.), *Global Political Economy: Contemporary Theories* (London: Routledge), pp. 197–214.

Pohjola, M. (2001) 'Information Technology and Economic Growth: A Cross Country Analysis', in Pohjola, M. (ed.), *Information Technology, Productivity and Economic Growth* (Oxford: Oxford University Press), pp. 242–56.

Poster, M. (1997) 'Cyberdemocracy: The Internet and the Public Sphere', in Holmes, D. (ed.), *Virtual Politics: Identity and Community in Cyberspace* (London: Sage Publications), pp. 212–28.

Prakash, S. (2001) 'TARAgyan: Empowerment through Education', *Development Alternatives Newsletter*, 11(7): 1–6.

Purkayastha, P. (2002) 'Skimming the cream', *Frontline*, 19(2), at http://www.frontlineonnet.com/fl1902/19020920.htm, accessed 15 November 2004.

Quah, D. T. (2001) 'The Weightless Economy in Economic Development', in Pohjola, M. (ed.), *Information Technology, Productivity, and Economic Growth: International Evidence and Implications for Economic Development* (Oxford: Oxford University Press), pp. 72–96.

Rai, S. (2003) 'Software Success Has India Worried', *New York Times*, 13 February, at http://www.nytimes.com/2003/02/13/business/worldbusiness/13RUPE.html, accessed 18 February 2003.

Raipuria, K. (2002) 'What Size the New Economy: A Conduit Approach', *Economic and Political Weekly*, 37(11): 16–22.

Ramachandraiah [one name only] (2003) 'Information Technology and Social Development', *Economic and Political Weekly*, 22–29 March: 1192–7.

Ramachandran, V. K. (1996) 'On Kerala's Development Achievements', in Drèze, J. and Sen, A. (eds), *Indian Development: Selected Regional Perspectives* (New Delhi: Oxford University Press), pp. 205–356.

——and Swaminathan, M. (2003) 'Introduction', in Ramachandran, V. K. and Swaminathan, M. (eds), *Agrarian Studies: Essays on Agrarian Relations in Less-Developed Countries* (New Delhi: Tulika Books), pp. xiii–xxviii.

——and Swaminathan, M. (2002) 'Rural Banking and Landless Labour Households: Institutional Reform and Rural Credit Markets in India', *Journal of Agrarian Change*, 2(4): 502–44.

Ramakumar, R. (2004) 'Socio-economic Characteristics of Agricultural Work: A Case Study of a Malabar Village', unpublished PhD thesis, Indian Statistical Institute, Kolkata.

Rasiah, R. and Best, M. (2000) *Industrial Transition in the Malaysian Electronics Industry* (Vienna: United Nations Industrial Development Organization).

Rawal, V. (2001) 'Agrarian Reform and Land Markets: A Study of Land Transactions in Two Villages of West Bengal, 1977–1995', *Economic Development and Cultural Change*, 49(3): 2537–44.

Reddy, N. K. and Graves, M. (2000) 'Electronic Support for Rural Healthcare Workers', in Bhatnagar and Schware, pp. 35–49.

Reddy, P. (1997) 'New Trends in Globalization of Corporate R&D and Implications for Innovation Capability in Host Countries: A Survey from India', *World Development*, 25: 1821–37.

Reich, R. B. (1991) *The Work of Nations: Preparing Ourselves for 21st Century Capitalism* (New York: Vintage).

Rifkin, J. (1995) *The End of Work: The Decline of the Global Labor Force and the Dawn of the Post-Market Era* (New York: G.P. Putnam's Sons).

Rodrik, D. (1992) 'The Limits of Trade Policy Reform in Developing Countries', *Journal of Economic Perspectives*, 6: 87–105.

Rosenberg, N. (1994) *Exploring the Black Box: Technology, Economics and History* (Cambridge: Cambridge University Press).

Rostow, W. W. (1960) *The Stages of Economic Growth: A Non-Communist Manifesto* (Cambridge: Cambridge University Press).

Sachs, J. (2001) 'A New Framework for Globalization', in *Efficiency, Equity, and Legitimacy: The Multilateral Trading System at the Millennium* (Washington, DC: Brookings Institution Press), pp. 63–80.

Sainath, P. (2003) 'The Bus to Mumbai', *The Hindu*, 15 June, at http://www.hinduonnet.com.

Salus, P. H. (1994) 'Unix at 25', *Byte*, 19(10): 75–82.

Sampson, G. P. and Snape, R. H. (1985) 'Identifying Issues in Trade in Services', *World Economy*, 8(2): 171–82.

Saxenian, A. (2002) 'The Silicon Valley Connection: Transnational Networks and Regional Development in Taiwan, China and India', *Science, Technology and Society*, 7: 117–49.

——(1999) *Silicon Valley's New Immigrant Entrepreneurs* (San Francisco: Public Policy Institute of California).

——Motoyama, Y. and Quan, X. (2002) *Local and Global Networks of Immigrant Professionals in Silicon Valley* (San Francisco: Public Policy Institute of California).

Schech, S. (2002) 'Wired for Change: The Links between ICTs and Development Discourses', *Journal of International Development*, 14: 13–23.

Schiller, D. (1999) *Digital Capitalism: Networking the Global Market System* (Cambridge, MA: MIT Press).

Schmitz, H. (1999) 'Collective Efficiency and Increasing Returns', *Cambridge Journal of Economics*, 23: 465–83.

Schneidewind, N. F. and Fenton, N. (1996) 'Do Standards Matter?', *IEEE Software*, 13(1): 22–4.

Schware, R. (1992) 'Software Industry Entry Strategies for Developing Countries: A "Walking on Two Legs" proposition', *World Development*, 20: 143–64.

——(1987) 'Software Industry in the Third World: Policy Guidelines, Institutional Options and Constraints', *World Development*, 15: 1249–67.

Sen, P. (1995) 'Indian Software Exports: An Assessment', *Economic and Political Weekly*, 30(7&8): 2053–8.

——(1994) 'Software Exports from India: A Systemic Analysis', *Electronics Information and Planning*, 22(2): 55–63.

Shapiro, C. and Varian, H. (1999) *Information Rules: A Strategic Guide to the Network Economy* (Boston: Harvard Business School Press).

Shiver, J. (2001) 'U.S. Tech Firms Abusing Visa Program, Critics Say', *Los Angeles Times*, 21 November.

Silicon India (2004a) 'India to Spend 16.7 B on Telecom: Gartner', 7 December, at www.siliconindia.com, accessed 8 December 2004.

——(2004b) 'Korean Firm Signs MoU for Semiconductor Facility', 6 December, at www.siliconindia.com, accessed 8 December 2004.

——(2004c) 'Nokia to Manufacture Cell Phones in India', 2 December, at www.siliconindia.com, accessed 8 December 2004.

Singh, S. (2003) 'India Calling', *Business World India*, at http://www.businessworldindia.com/Nov1003/coverstory02.asp, accessed 20 February 2005.

Sneed, H. M. (1995) 'Planning the Reengineering of Legacy Systems', *IEEE Software*, 12(1): 24–34.

Sood, A. D. (2003) 'Information Nodes in Rural Landscape', *Information for Development*, 1(1): 14–21.

Sparks, C. (1994) 'Civil Society and Information Society as Guarantors of Progress', in Splichel, S., Calabrese, A. and Spark, C. (eds), *Information Society and Civil Society: Perspectives on the Changing World Order* (West Lafayette, IN: Perdue University Press), pp. 21–49.

Sreekumar, T. T. (2003) 'De-hyping ICTs: ICT Innovations by Civil Society Organizations in Rural India', *Information for Development*, 1(1): 22–7.

Sridharan, E. (1996) *The Political Economy of Industrial Promotion: Indian, Brazilian, and Korean Electronics in Comparative Perspective 1969–1994* (London: Praeger).

Srinivas, A. and Jayashankar, M. (2002) 'Hard Times, Hard Lessons: The New Software Solutions', *Business World*, 25 February: 24–9.

Srinivas, M. (1996) *Indian Society through Personal Writings* (Delhi: Oxford University Press).

Steen, M. (2001) 'Demand for IT workers is down 44 per cent', *San Jose Mercury News*, 2 April.

Stehr, N. (1994) *Knowledge societies* (London: Sage).
Steinmueller, W. E. (1996) 'The U.S. Software Industry: An Analysis and Interpretive History', in Mowery, D. C. (ed.), *The International Computer Software Industry: A Comparative Study of Industry Evolution and Structure* (New York: Oxford University Press), pp. 15–52.
Stiroh, K. J. (2002) 'Information Technology and the U.S. Productivity Revival: What do the Industry Data Say?', *American Economic Review*, 92: 1559–76.
Stoll, C. (1995) *Silicon Snake Oil: Second Thoughts on the Information Highway* (New York: Doubleday).
Subramanian, C. R. (1992) *India and the Computer: A Study of Planned Development* (New Delhi: Oxford University Press).
Sutcliffe, B. (2004) 'World Inequality and Globalization', *Oxford Review of Economic Policy*, 20: 15–37.
——(1999) 'The Place of Development in Theories of Imperialism and Globalization', in Munck, R. and O'Hearn, D. (eds), *Critical Development Theory: Contributions to a New Paradigm* (London: Zed Books), pp. 135–54.
Tempest, R. (1996) 'Barbie and the World Economy', *Los Angeles Times*, 22 September: A1 and A12.
Theil, S. (2002) *Quangos: Trends, Causes and Consequences* (Aldershot: Ashgate).
Thomas, J. J. (2005a) 'Kerala's Industrial Backwardness: A Case of Path Dependence in Industrialization?', *World Development*, 33(5): 763–83.
——(2005b) 'New Technologies for India's Development', in Parikh, K. S. and Radhakrishna, R. (eds), *India Development Report 2004–05* (New Delhi: Oxford University Press, pp. 126–40.
Thompson, G. (2002) 'Free-Market Upheaval Grinds Mexico's Middle Class', *New York Times*, 4 September, at http://www.nytimes.com/2002/09/04/international/americas/04MEXI, accessed 4 September 2002.
Times of India (2001a) 'Dotcoms See Hope Despite Gloom', 30 December.
——(2001b) 'VCs Backing Off After Tech Burst', 16 August.
Touraine, A. (1969) *La Société Post-Industriale* (Paris: Denoel).
Udell, J. (1993) 'India's Software Edge', *Byte*, 18(10): 55–60.
UNCTAD (2003) *E-Commerce and Development Report 2003* (Geneva: UNCTAD).
UNDP (2002) *Human Development Report 2002: Deepening Democracy in Fragmented World* (New York: Oxford University Press).
——(2001) *Human Development Report 2001: Making New Technologies Work for Human Development* (New York: Oxford University Press).
——(1999) *Human Development Report 1999: Globalization with a Human Face* (New York: Oxford University Press).
United Nations (1999) *Economic and Social Survey of the Asia and the Pacific 1999* (New York: UN Publications).
Urry, J. (2000) *Sociology beyond Societies: Mobilities for the Twenty-First Century* (London and New York: Routledge).
Usui, N. (1996) 'Policy Adjustment to the Oil Boom and their Evaluation: The Dutch Disease in Indonesia', *World Development*, 24(5): 887–900.
Vaas, L. (2001) 'Testing the New IT Job Pool', *eWEEK*, 21 May.
Vahid, F. (2003) 'The Softening of Hardware', *IEEE Computer*, 36(4): 27–34.
Van Wijnbergen, S. J. G. (1984) 'Inflation, Employment and the Dutch Disease in Oil Exporting Countries: A Short Run Disequilibrium Analysis', *Quarterly Journal of Economics*, 19: 233–50.

Vasagam, R. M. (1999) 'Internet Kiosks: Tamil Nadu Leads the Way', *Voices for Change*, 3(3): 23–4.
Velden, M. V. (2002) 'Knowledge Facts, Knowledge Fiction: The Role of ICTs in Knowledge Management for Development', *Journal of International Development*, 14: 25–37.
Verma, S. (2000a) 'The IT Task Force: Two years down', *Dataquest*, 18(20): 117–24.
——(2000b) 'The IT Task Force Scorecard II: Bridging the Digital Divide', *Dataquest*, 18(21): 70–8.
——(2000c) 'The IT Task Force Scorecard III: The IT Industry Impact', *Dataquest*, 18(22): 111–20.
Vijayaditya, N. (2000) 'A Wired Village: The Warana Experiment', in Bhatnagar and Schware, pp. 132–40.
Virmani, A. (2004) 'Economic Growth, Governance and Voting Behaviour: An Application To Indian Elections', working paper no. 138, Indian Council For Research on International Economic Relations, New Delhi, at http://www.icrier.org, accessed 14 December 2004.
von Hippel, E. (1988) *The Sources of Innovation* (New York: Oxford University Press).
Wade, R. H. (2002) 'Bridging the Digital Divide: New Route to Development or New Form of Dependency?', *Global Governance*, 8(4): 53–85.
WIDER (2000) *World Income Inequality Database* (Helsinki: World Institute for Development Economics Research), at http://www.wider.unu.edu/wiid, accessed 21 November 2003.
Wolf, W. (2002) 'What is Embedded Computing?', *IEEE Computer*, 35(1): 136–7.
Wolff, E. N. (2002) 'The Impact of IT Investment on Income and Wealth Inequality in the Postwar US Economy', *Information Economics and Policy*, 14: 233–51.
Wong, P. K. (2001) 'The Contribution of Information Technology to the Rapid Economic Growth of Singapore', in Pohjola, M. (ed.), *Information Technology, Productivity and Economic Growth* (Oxford: Oxford University Press), pp. 221–41.
Wood, A. (1995) 'How Trade Hurts Unskilled Workers', *Journal of Economic Perspectives*, 9: 57–80.
World Bank (2004) *Unlocking Andhra Pradesh's Growth Potential: An Agenda to Achieve the Vision 2020 Growth Targets* (Washington, DC: World Bank), at www.worldbank.org, accessed 14 December 2004.
——(2003a) *India: Sustaining Reform, Reducing Poverty – A World Bank Development Policy Review* (Washington, DC: World Bank), at www.worldbank.org, accessed 14 December 2004.
——(2003b) *World Development Report 2003* (Washington, DC: World Bank).
——(2000) 'India: Scientific and Technical Manpower Development in India', (Washington, DC: World Bank, Education Sector Unit, South Asia Region), report no. 20416-IN.
——(1999) *World Development Report 1998–99: Knowledge for Development* (Oxford and New York: Oxford University Press).
World Trade Organisation (2003) *International Trade Statistics* (Geneva: WTO).

Index

Note: 'n.' after a page reference indicates the number of a note on that page.